Student Solutions Manual

Fundamentals of Analytical Chemistry

NINTH EDITION

Douglas A. Skoog
Stanford University

Donald M. West
San Jose State University

F. James Holler
University of Kentucky

Stanley R. Crouch
Michigan State University

Prepared by

Stanley R. Crouch
Michigan State University

F. James Holler
University of Kentucky

BROOKS/COLE
CENGAGE Learning

Australia • Brazil • Japan • Korea • Mexico • Singapore • Spain • United Kingdom • United States

ISBN-13: 978-0-495-55834-7
ISBN-10: 0-495-55834-6

Brooks/Cole
20 Davis Drive
Belmont, CA 94002-3098
USA

Cengage Learning is a leading provider of customized learning solutions with office locations around the globe, including Singapore, the United Kingdom, Australia, Mexico, Brazil, and Japan. Locate your local office at: **www.cengage.com/global**

Cengage Learning products are represented in Canada by Nelson Education, Ltd.

To learn more about Brooks/Cole, visit **www.cengage.com/brookscole**

Purchase any of our products at your local college store or at our preferred online store **www.cengagebrain.com**

Printed in the United States of America
1 2 3 4 5 6 7 16 15 14 13 12

Table of Contents

Chapter 3

3-1.   **(a)** SQRT returns the square root of a number or result of a calculation.

(b) AVERAGE returns the arithmetic mean of a series of numbers.

(c) PI returns the value of pi accurate to 15 digits

(d) FACT returns the factorial of a number, equal to $1 \times 2 \times 3 \times \ldots \times$ number.

(e) EXP returns e raised to the value of a given number.

(f) LOG returns the logarithm of a number to a base specified by the user.

Chapter 4

4-1.   **(a)** The *millimole* is an amount of a chemical species, such as an atom, an ion, a

molecule or an electron. There are

$$6.02 \times 10^{23} \frac{\text{particles}}{\text{mole}} \times 10^{-3} \frac{\text{mole}}{\text{millimole}} = 6.02 \times 10^{20} \frac{\text{particles}}{\text{millimole}}$$

(c) The *millimolar mass* is the mass in grams of one millimole of a chemical species.

4-3.   The liter: $1\,L = \dfrac{1000\,\text{mL}}{1\,L} \times \dfrac{1\,\text{cm}^3}{1\,\text{mL}} \times \left(\dfrac{1\,\text{m}}{100\,\text{cm}}\right)^3 = 10^{-3}\,\text{m}^3$

Molar concentration: $1\,M = \dfrac{1\,\text{mol}}{1\,L} \times \dfrac{1\,L}{10^{-3}\,\text{m}^3} = \dfrac{1\,\text{mol}}{10^{-3}\,\text{m}^3}$

4-4.   **(a)** $3.2 \times 10^8\,\text{Hz} \times \dfrac{1\,\text{MHz}}{10^6\,\text{Hz}} = 320\,\text{MHz}$

(c) $8.43 \times 10^7\,\mu\text{mol} \times \dfrac{1\,\text{mol}}{10^6\,\mu\text{mol}} = 84.3\,\text{mol}$

(e) $8.96 \times 10^6\,\text{nm} \times \dfrac{1\,\text{mm}}{10^6\,\text{nm}} = 8.96\,\text{mm}$

4-5.   For oxygen, for example 15.999 u/atom = 15.999 g/6.022 $\times 10^{23}$ atoms = 15.999 g/mol.

So 1 u = 1 g/mol.

Thus, 1g = 1mol u.

4-7. $2.92\,\text{g Na}_3\text{PO}_4 \times \dfrac{1\,\text{mol Na}_3\text{PO}_4}{163.94\,\text{g}} \times \dfrac{3\,\text{mol Na}^+}{1\,\text{mol Na}_3\text{PO}_4} \times \dfrac{6.022 \times 10^{23}\,\text{Na}^+}{1\,\text{mol Na}^+} = 3.22 \times 10^{22}\,\text{Na}^+$

4-9.   **(a)** $8.75 \text{ g B}_2\text{O}_3 \times \dfrac{2 \text{ mol B}}{1 \text{ mol B}_2\text{O}_3} \times \dfrac{1 \text{ mol B}_2\text{O}_3}{69.62 \text{ g B}_2\text{O}_3} = 0.251 \text{ mol B}$

(b) $167.2 \text{ mg Na}_2\text{B}_4\text{O}_7 \bullet 10\text{H}_2\text{O} \times \dfrac{1 \text{ g}}{1000 \text{ mg}} \times \dfrac{7 \text{ mol O}}{1 \text{ mol Na}_2\text{B}_4\text{O}_7 \bullet 10\text{H}_2\text{O}}$

$\times \dfrac{1 \text{ mol Na}_2\text{B}_4\text{O}_7 \bullet 10\text{H}_2\text{O}}{381.37 \text{ g}} = 3.07 \times 10^{-3} \text{ mol O} = 3.07 \text{ mmol}$

(c) $4.96 \text{ g Mn}_3\text{O}_4 \times \dfrac{1 \text{ mol Mn}_3\text{O}_4}{228.81 \text{ g Mn}_3\text{O}_4} \times \dfrac{3 \text{ mol Mn}}{1 \text{ mol Mn}_3\text{O}_4} = 6.50 \times 10^{-2} \text{ mol Mn}$

(d) $333 \text{ mg CaC}_2\text{O}_4 \times \dfrac{1 \text{ g}}{1000 \text{ mg}} \times \dfrac{\text{mol CaC}_2\text{O}_4}{128.10 \text{ g CaC}_2\text{O}_4} \times \dfrac{2 \text{ mol C}}{1 \text{ mol CaC}_2\text{O}_4} = 5.20 \times 10^{-3} \text{ mol C}$

$= 5.20 \text{ mmol}$

4-11.   **(a)** $\dfrac{0.0555 \text{ mol KMnO}_4}{\text{L}} \times \dfrac{1000 \text{ mmol}}{1 \text{ mol}} \times 2.00 \text{ L} = 111 \text{ mmol KMnO}_4$

(b) $\dfrac{3.25 \times 10^{-3} \text{ M KSCN}}{\text{L}} \times \dfrac{1000 \text{ mmol}}{1 \text{ mol}} \times \dfrac{\text{L}}{1000 \text{ mL}} \times 750 \text{ mL}$

$= 2.44 \text{ mmol KSCN}$

(c) $\dfrac{3.33 \text{ mg CuSO}_4}{1 \text{ L}} \times \dfrac{1 \text{ g}}{1000 \text{ mg}} \times \dfrac{1 \text{ mol CuSO}_4}{159.61 \text{ g CuSO}_4} \times \dfrac{1000 \text{ mmol}}{1 \text{ mol}} \times 3.50 \text{ L}$

$= 7.30 \times 10^{-2} \text{ mmol CuSO}_4$

(d) $\dfrac{0.414 \text{ mol KCl}}{1 \text{ L}} \times \dfrac{1000 \text{ mmol}}{1 \text{ mol}} \times \dfrac{1 \text{ L}}{1000 \text{ mL}} \times 250 \text{ mL} = 103.5 \text{ mmol KCl}$

4-13.   **(a)** $0.367 \text{ mol HNO}_3 \times \dfrac{63.01 \text{ g HNO}_3}{1 \text{ mol HNO}_3} \times \dfrac{1000 \text{ mg}}{1 \text{ g}} = 2.31 \times 10^4 \text{ mg HNO}_3$

(b) $245 \text{ mmol MgO} \times \dfrac{1 \text{ mol}}{1000 \text{ mmol}} \times \dfrac{40.30 \text{ g MgO}}{1 \text{ mol MgO}} \times \dfrac{1000 \text{ mg}}{1 \text{ g}} = 9.87 \times 10^3 \text{ mg MgO}$

(c) $12.5 \text{ mol NH}_4\text{NO}_3 \times \dfrac{80.04 \text{ g NH}_4\text{NO}_3}{1 \text{ mol NH}_4\text{NO}_3} \times \dfrac{1000 \text{ mg}}{1 \text{ g}} = 1.00 \times 10^6 \text{mg NH}_4\text{NO}_3$

(d) $4.95 \text{ mol (NH}_4)_2\text{Ce(NO}_3)_6 \times \dfrac{548.23 \text{ g (NH}_4)_2\text{Ce(NO}_3)_6}{1 \text{ mol (NH}_4)_2\text{Ce(NO}_3)_6} \times \dfrac{1000 \text{ mg}}{1 \text{ g}}$

$= 2.71 \times 10^6 \text{mg (NH}_4)_2\text{Ce(NO}_3)_6$

4-15. **(a)** $\dfrac{0.350 \text{ mol sucrose}}{1 \text{ L}} \times \dfrac{1 \text{ L}}{1000 \text{ mL}} \times \dfrac{342 \text{ g sucrose}}{1 \text{ mol sucrose}} \times \dfrac{1000 \text{ mg}}{1 \text{ g}}$

$\times 16.0 \text{ mL} = 1.92 \times 10^3 \text{mg sucrose}$

(b) $\dfrac{3.76 \times 10^{-3} \text{mol H}_2\text{O}_2}{1 \text{ L}} \times \dfrac{34.02 \text{ g H}_2\text{O}_2}{1 \text{ mol H}_2\text{O}_2} \times \dfrac{1000 \text{ mg}}{1 \text{ g}}$

$\times 1.92 \text{ L} = 246 \text{ mg H}_2\text{O}_2$

4-16. **(a)** $\dfrac{0.264 \text{ mol H}_2\text{O}_2}{1 \text{ L}} \times \dfrac{1 \text{ L}}{1000 \text{ mL}} \times \dfrac{34.02 \text{ g H}_2\text{O}_2}{1 \text{ mol H}_2\text{O}_2} \times 250 \text{ mL}$

$= 2.25 \text{ g H}_2\text{O}_2$

(b) $\dfrac{5.75 \times 10^{-4} \text{mol benzoic acid}}{1 \text{ L}} \times \dfrac{1 \text{ L}}{1000 \text{ mL}} \times \dfrac{122 \text{ g benzoic acid}}{1 \text{ mol benzoic acid}}$

$\times 37.0 \text{ mL} = 2.60 \times 10^{-3} \text{g benzoic acid}$

4-17. **(a)** $\text{pNa} = -\log(0.0635 + 0.0403) = -\log(0.1038) = 0.9838$

$\text{pCl} = -\log(0.0635) = 1.197$

$\text{pOH} = -\log(0.0403) = 1.395$

(c)

$\text{pH} = -\log(0.400) = 0.398$

$\text{pCl} = -\log(0.400 + 2 \times 0.100) = -\log(0.600) = 0.222$

$\text{pZn} = -\log(0.100) = 1.00$

4

(e)

$$pK = -\log(4 \times 1.62 \times 10^{-7} + 5.12 \times 10^{-7}) = -\log(1.16 \times 10^{-6}) = 5.936$$

$$pOH = -\log(5.12 \times 10^{-7}) = 6.291$$

$$pFe(CN)_6 = -\log(1.62 \times 10^{-7}) = 6.790$$

4-18. **(a)** $pH = 4.31$, $\log[H_3O^+] = -4.31$, $[H_3O^+] = 4.9 \times 10^{-5}$ M

as in part (a)

(c) $[H_3O^+] = 0.26$ M

(e) $[H_3O^+] = 2.4 \times 10^{-8}$ M

(g) $[H_3O^+] = 5.8$ M

4-19. **(a)** $pNa = pBr = -\log(0.0300) = 1.523$

(c) $pBa = -\log(5.5 \times 10^{-3}) = 2.26$; $pOH = -\log(2 \times 5.5 \times 10^{-3}) = 1.96$

(e) $pCa = -\log(8.7 \times 10^{-3}) = 2.06$; $pBa = -\log(6.6 \times 10^{-3}) = 2.18$

  $pCl = -\log(2 \times 8.7 \times 10^{-3} + 2 \times 6.6 \times 10^{-3}) = -\log(0.0306) = 1.51$

4-20. **(a)** $pH = 1.020$; $\log[H_3O^+] = -1.020$; $[H_3O^+] = 0.0955$ M

(c) $pBr = 7.77$; $[Br^-] = 1.70 \times 10^{-8}$ M

(e) $pLi = 12.35$; $[Li^+] = 4.5 \times 10^{-13}$ M

(g) $pMn = 0.135$; $[Mn^{2+}] = 0.733$ M

4-21. **(a)** $1.08 \times 10^3 \, ppm \, Na^+ \times \dfrac{1}{10^6 \, ppm} \times \dfrac{1.02 \, g}{1 \, mL} \times \dfrac{1000 \, mL}{1 \, L} \times \dfrac{1 \, mol \, Na^+}{22.99 \, g} = 4.79 \times 10^{-2} \, M \, Na^+$

  $270 \, ppm \, SO_4^{2-} \times \dfrac{1}{10^6 \, ppm} \times \dfrac{1.02 \, g}{1 \, mL} \times \dfrac{1000 \, mL}{1 \, L} \times \dfrac{1 \, mol \, SO_4^{3-}}{96.06 \, g} = 2.87 \times 10^{-3} \, M \, SO_4^{2-}$

(b) $pNa = -\log(4.79 \times 10^{-2}) = 1.320$

  $pSO_4 = -\log(2.87 \times 10^{-3}) = 2.542$

5

4-23. (a)

$$\frac{5.76 \text{ g KCl} \cdot \text{MgCl}_2 \cdot 6\text{H}_2\text{O}}{2.00 \text{ L}} \times \frac{1 \text{ mol KCl} \cdot \text{MgCl}_2 \cdot 6\text{H}_2\text{O}}{277.85 \text{ g}} = 1.04 \times 10^{-2} \text{M KCl} \cdot \text{MgCl}_2 \cdot 6\text{H}_2\text{O}$$

(b) There is 1 mole of Mg^{2+} per mole of $KCl \cdot MgCl_2$, so the molar concentration of Mg^{2+}

is the same as the molar concentration of $KCl \cdot MgCl_2$ or $1.04 \times 10^{-2} M$

(c) $1.04 \times 10^{-2} \text{M KCl} \cdot \text{MgCl}_2 \cdot 6\text{H}_2\text{O} \times \dfrac{3 \text{ mol Cl}^-}{1 \text{ mol KCl} \cdot \text{MgCl}_2 \cdot 6\text{H}_2\text{O}} = 3.12 \times 10^{-2} \text{M Cl}^-$

(d) $\dfrac{5.76 \text{ g KCl} \cdot \text{MgCl}_2 \cdot 6\text{H}_2\text{O}}{2.00 \text{ L}} \times \dfrac{1 \text{ L}}{1000 \text{ mL}} \times 100\% = 0.288\% \text{ (w/v)}$

(e) $\dfrac{3.12 \times 10^{-2} \text{ mol Cl}^-}{1 \text{ L}} \times \dfrac{1 \text{ L}}{1000 \text{ mL}} \times \dfrac{1000 \text{ mmol}}{1 \text{ mol}} \times 25 \text{ mL} = 7.8 \times 10^{-1} \text{ mmol Cl}^-$

(f)
$$1.04 \times 10^{-2} \text{ M KCl} \cdot \text{MgCl}_2 \cdot 6\text{H}_2\text{O} \times \frac{1 \text{ mol K}^+}{1 \text{ mol KCl} \cdot \text{MgCl}_2 \cdot 6\text{H}_2\text{O}} \times \frac{39.10 \text{ g K}^+}{1 \text{ mol K}^+} \times \frac{1000 \text{ mg}}{1 \text{ g}}$$

$$= \frac{407 \text{ mg}}{1 \text{ L}} = 407 \text{ ppm K}^+$$

(g) $pMg = -\log(1.04 \times 10^{-2}) = 1.983$

(h) $pCl = -\log(3.12 \times 10^{-2}) = 1.506$

4-25. (a) $6.42\% \text{ Fe(NO}_3)_3 = \dfrac{6.42 \text{ g Fe(NO}_3)_3}{100 \text{ g solution}} \times \dfrac{1.059 \text{ g}}{\text{mL}} \times \dfrac{1000 \text{ mL}}{1 \text{ L}} \times \dfrac{1 \text{ mol Fe(NO}_3)_3}{241.86 \text{ g}}$

$= 2.81 \times 10^{-1} \text{ M Fe(NO}_3)_3 = 0.281 \text{ M}$

(b)

$$2.81 \times 10^{-1} \text{M Fe(NO}_3)_3 = \frac{2.81 \times 10^{-1} \text{ mol Fe(NO}_3)_3}{\text{L}} \times \frac{3 \text{ mol NO}_3^-}{1 \text{ mol Fe(NO}_3)_3} = 8.43 \times 10^{-1} \text{M NO}_3^-$$

(c) $\dfrac{2.81 \times 10^{-1} \text{ mol Fe(NO}_3)_3}{\text{L}} \times \dfrac{241.86 \text{ g Fe(NO}_3)_3}{1 \text{ mol}} \times 1 \text{ L} = 6.80 \times 10^1 \text{ g Fe(NO}_3)_3 = 68.0 \text{ g}$

4-27. **(a)** $\dfrac{4.75 \text{ g C}_2\text{H}_5\text{OH}}{100 \text{ mL soln}} \times 500 \text{ mL soln} = 2.38 \times 10^1 \text{ g C}_2\text{H}_5\text{OH}$

Weigh 23.8 g ethanol and add enough water to give a final volume of 500 mL

$$4.75\% \text{ (w/w) C}_2\text{H}_5\text{OH} = \dfrac{4.75 \text{ g C}_2\text{H}_5\text{OH}}{100 \text{ g soln}} \times 500 \text{ g soln} = 2.38 \times 10^1 \text{ g C}_2\text{H}_5\text{OH}$$

(b) 500 g soln = 23.8 g C$_2$H$_5$OH + x g water

x g water = 500 g soln $-$ 23.8 g C$_2$H$_5$OH = 476.2 g water

Mix 23.8 g ethanol with 476.2 g water

$$4.75\% \text{ (v/v) C}_2\text{H}_5\text{OH} = \dfrac{4.75 \text{ mL C}_2\text{H}_5\text{OH}}{100 \text{ mL soln}}$$

(c) $\dfrac{4.75 \text{ mL C}_2\text{H}_5\text{OH}}{100 \text{ mL soln}} \times 500 \text{ mL soln} = 2.38 \times 10^1 \text{ mL C}_2\text{H}_5\text{OH}$

Dilute 23.8 mL ethanol with enough water to give a final volume of 500 mL.

4-29.

$$\dfrac{6.00 \text{ mol H}_3\text{PO}_4}{L} \times \dfrac{1 \text{ L}}{1000 \text{ mL}} \times 750 \text{ mL} = 4.50 \text{ mol H}_3\text{PO}_4$$

$$\dfrac{86 \text{ g H}_3\text{PO}_4}{100 \text{ g reagent}} \times \dfrac{1.71 \text{ g reagent}}{\text{g water}} \times \dfrac{\text{g water}}{\text{mL}} \times \dfrac{1000 \text{ mL}}{L} \times \dfrac{\text{mol H}_3\text{PO}_4}{98.0 \text{ g}}$$

$$= \dfrac{1.50 \times 10^1 \text{ mol H}_3\text{PO}_4}{L}$$

volume 86% (w/w) H$_3$PO$_4$ required = $4.50 \text{ mol H}_3\text{PO}_4 \times \dfrac{L}{1.50 \times 10^1 \text{ mol H}_3\text{PO}_4} = 3.00 \times 10^{-1} \text{ L}$

$$0.0750 \text{ M AgNO}_3 = \dfrac{0.0750 \text{ mol AgNO}_3}{L}$$

4-31. **(a)** $= \dfrac{0.0750 \text{ mol AgNO}_3}{L} \times \dfrac{169.87 \text{ g AgNO}_3}{1 \text{ mol}} \times \dfrac{1 \text{ L}}{1000 \text{ mL}} \times 500 \text{ mL}$

$= 6.37 \text{ g AgNO}_3$

Dissolve 6.37 g AgNO$_3$ in enough water to give a final volume of 500 mL.

(b)

$$\frac{0.285 \text{ mol HCl}}{L} \times 1 \text{ L} = 0.285 \text{ mol HCl}$$

$$0.285 \text{ mol HCl} \times \frac{1 \text{ L}}{6.00 \text{ mol HCl}} = 4.75 \times 10^{-2} \text{ L HCl}$$

Take 47.5 mL of the 6.00 M HCl and dilute to 1.00 L with water.

(c)

$$\frac{0.0810 \text{ mol K}^+}{L} \times \frac{1 \text{ L}}{1000 \text{ mL}} \times 400 \text{ mL} = 3.24 \times 10^{-2} \text{ mol K}^+$$

$$3.24 \times 10^{-2} \text{ mol K}^+ \times \frac{1 \text{ mol K}_4\text{Fe(CN)}_6}{4 \text{ mol K}^+} \times \frac{368.43 \text{ g K}_4\text{Fe(CN)}_6}{\text{mol}} = 2.98 \text{ g K}_4\text{Fe(CN)}_6$$

Dissolve 2.98 g $K_4Fe(CN)_6$ in enough water to give a final volume of 400 mL.

(d)

$$\frac{3.00 \text{ g BaCl}_2}{100 \text{ mL soln}} \times 600 \text{ mL} = 1.8 \times 10^1 \text{g BaCl}_2$$

$$1.8 \times 10^1 \text{g BaCl}_2 \times \frac{1 \text{ mol BaCl}_2}{208.23 \text{ g}} \times \frac{L}{0.400 \text{ mol BaCl}_2} = 2.16 \times 10^{-1} \text{ L}$$

Take 216 mL of the 0.400 M $BaCl_2$ solution and dilute to 600 mL with water

(e)

$$\frac{0.120 \text{ mol HClO}_4}{L} \times 2.00 \text{ L} = 0.240 \text{ mol HClO}_4$$

$$\frac{71 \text{ g HClO}_4}{100 \text{ g reagent}} \times \frac{1.67 \text{ g reagent}}{1 \text{ g water}} \times \frac{1 \text{ g water}}{1 \text{ mL}} \times \frac{1000 \text{ mL}}{1 \text{ L}} \times \frac{\text{mol HClO}_4}{100.46 \text{ g}}$$

$$= \frac{1.18 \times 10^1 \text{mol HClO}_4}{L}$$

volume 71% (w/w) $HClO_4$ required = $0.240 \text{ mol HClO}_4 \times \frac{L}{1.18 \times 10^1 \text{mol HClO}_4} = 2.03 \times 10^{-2} \text{ L}$

Take 20.3 mL of the concentrated reagent and dilute to 2.00 L with water.

$$60 \text{ ppm Na}^+ = \frac{60 \text{ mg Na}^+}{\text{L soln}}$$

$$\frac{60 \text{ mg Na}^+}{\text{L soln}} \times 9.00 \text{ L} = 5.4 \times 10^2 \text{mg Na}^+$$

(f)

$$5.4 \times 10^2 \text{mg Na}^+ \times \frac{1 \text{ g}}{1000 \text{ mg}} \times \frac{1 \text{ mol Na}^+}{22.99 \text{ g}} = 2.35 \times 10^{-2} \text{ mol Na}^+$$

$$2.35 \times 10^{-2} \text{ mol Na}^+ \times \frac{1 \text{ mol Na}_2\text{SO}_4}{2 \text{ mol Na}^+} \times \frac{142.04 \text{ g Na}_2\text{SO}_4}{1 \text{ mol}} = 1.7 \text{ g Na}_2\text{SO}_4$$

Dissolve 1.7 g Na_2SO_4 in enough water to give a final volume of 9.00 L.

4-33.

$$\frac{0.250 \text{ mol La}^{3+}}{\text{L}} \times \frac{1 \text{ L}}{1000 \text{ mL}} \times 50.0 \text{ mL} = 1.25 \times 10^{-2} \text{ mol La}^{3+}$$

$$0.302 \text{ M IO}_3^- = \frac{0.302 \text{ mol IO}_3^-}{1 \text{ L}} \times \frac{1 \text{ L}}{1000 \text{ mL}} \times 75.0 \text{ mL} = 2.27 \times 10^{-2} \text{ mol IO}_3^-$$

Because each mole of $La(IO_3)_3$ requires three moles IO_3^-, IO_3^- is the limiting reagent.

Thus,

$$2.27 \times 10^{-2} \text{ mol IO}_3^- \times \frac{1 \text{ mol La(IO}_3)_3}{3 \text{ mol IO}_3^-} \times \frac{663.6 \text{ g La(IO}_3)_3}{1 \text{ mol}} = 5.01 \text{ g La(IO}_3)_3 \text{formed}$$

4-35. A balanced chemical equation can be written as:

$$Na_2CO_3 + 2HCl \rightarrow 2NaCl + H_2O + CO_2(g)$$

(a)

$$0.2220 \text{ g Na}_2\text{CO}_3 \times \frac{1 \text{ mol Na}_2\text{CO}_3}{105.99 \text{ g}} = 2.094 \times 10^{-3} \text{mol Na}_2\text{CO}_3$$

$$\frac{0.0731 \text{ mol HCl}}{\text{L}} \times \frac{1 \text{ L}}{1000 \text{ mL}} \times 100.0 \text{ mL} = 7.31 \times 10^{-3} \text{ mol HCl}$$

Because one mole of CO_2 is evolved for every mole Na_2CO_3 reacted, Na_2CO_3 is the

limiting reagent. Thus

$$2.094 \times 10^{-3} \text{ mol } Na_2CO_3 \times \frac{1 \text{ mol } CO_2}{1 \text{ mol } Na_2CO_3} \times \frac{44.00 \text{ g } CO_2}{1 \text{ mol}} = 9.214 \times 10^{-2} \text{ g } CO_2 \text{ evolved}$$

(b)

$$\text{amnt HCl left} = 7.31 \times 10^{-3} \text{ mol} - (2 \times 2.094 \times 10^{-3} \text{ mol}) = 3.12 \times 10^{-3} \text{ mol}$$

$$\frac{3.12 \times 10^{-3} \text{ mol HCl}}{100.0 \text{ mL}} \times \frac{1000 \text{ mL}}{1 \text{ L}} = 3.12 \times 10^{-2} \text{ M HCl}$$

4-37　A balanced chemical equation can be written as:

$$Na_2SO_3 + 2HClO_4 \rightarrow 2NaClO_4 + H_2O + SO_2(g)$$

(a)

$$0.3132 \text{ M } Na_2SO_3 = \frac{0.3132 \text{ mol } Na_2SO_3}{L} \times \frac{L}{1000 \text{ mL}} \times 75 \text{ mL} = 2.3 \times 10^{-2} \text{ mol } Na_2SO_3$$

$$0.4025 \text{ M } HClO_4 = \frac{0.4025 \text{ mol } HClO_4}{L} \times \frac{L}{1000 \text{ mL}} \times 150.0 \text{ mL} = 6.038 \times 10^{-2} \text{ mol } HClO_4$$

Because one mole SO_2 is evolved per mole Na_2SO_3, Na_2SO_3 is the limiting reagent.

Thus,

$$2.3 \times 10^{-2} \text{ mol } Na_2SO_3 \times \frac{\text{mol } SO_2}{\text{mol } Na_2SO_3} \times \frac{64.06 \text{ g } SO_2}{\text{mol}} = 1.5 \text{ g } SO_2 \text{ evolved}$$

(b)

$$\text{mol } HClO_4 \text{ unreacted} = (6.038 \times 10^{-2} \text{ mol} - (2 \times 2.3 \times 10^{-2}) = 1.4 \times 10^{-2} \text{ mol}$$

$$\frac{1.4 \times 10^{-2} \text{ mol } HClO_4}{225 \text{ mL}} \times \frac{1000 \text{ mL}}{L} = 6.4 \times 10^{-2} \text{ M } HClO_4 = 0.064 \text{ M}$$

4-39. A balanced chemical equation can be written as:

$$AgNO_3 + KI \rightarrow AgI(s) + KNO_3$$

$$24.31 \text{ ppt KI} \times \frac{1}{10^3 \text{ppt}} \times \frac{1 \text{ g}}{1 \text{ mL}} \times 200.0 \text{ mL} \times \frac{1 \text{ mol KI}}{166.0 \text{ g}} = 2.93 \times 10^{-2} \text{ mol KI}$$

$$2.93 \times 10^{-2} \text{ mol KI} \times \frac{1 \text{ mol AgNO}_3}{1 \text{ mol KI}} \times \frac{1 \text{ L}}{0.0100 \text{ mol AgNO}_3} = 2.93 \text{ L AgNO}_3$$

2.93 L of 0.0100 M $AgNO_3$ would be required to precipitate I^- as AgI.

Chapter 5

5-1.   **(a)** Random error causes data to be scattered more or less symmetrically around a mean

value while systematic error causes the mean of a data set to differ from the accepted

value.

(c) The absolute error of a measurement is the difference between the measured value and

the true value while the relative error is the absolute error divided by the true value.

5-2.   (1) Meter stick slightly longer or shorter than 1.0 m – systematic error.

(2) Markings on the meter stick always read from a given angle – systematic error.

(3) Variability in the sequential movement of the 1-m metal rule to measure the full 3-m

table width – random error.

(4) Variability in interpolation of the finest division of the meter stick – random error.

5-4.   (1) The analytical balance is miscalibrated.

(2) After weighing an empty vial, fingerprints are placed on the vial while adding sample

to the vial.

(3) A hygroscopic sample absorbs water from the atmosphere while placing it in a

weighing vial.

5-5.   (1) The pipet is miscalibrated and holds a slightly different volume of liquid than the

indicated volume.

(2) The user repetitively reads the volume marking on the pipet from an angle rather than

at eye level.

(3) The inner surfaces of the pipet are contaminated.

5-7. Both constant and proportional systematic errors can be detected by varying the sample size. Constant errors do not change with the sample size while proportional errors increase or decrease with increases or decreases in the samples size.

5-8. **(a)** $(-0.4 \text{ mg}/500 \text{ mg}) \times 100\% = -0.08\%$

As in part (a)

(c) -0.27%

5-9. **(a)** First determine how much gold is needed to achieve the desired relative error.

$(-0.4 \text{ mg}/-0.1\%) \times 100\% = 400 \text{ mg gold}$

Then determine how much ore is needed to yield the required amount of gold.

$(400 \text{ mg}/1.2\%) \times 100\% = 33{,}000 \text{ mg ore}$ **or** 33 g ore

(c) 4.2 g ore

5-10 **(a)** $(0.03/50.00) \times 100\% = 0.060\%$

As in part (a)

(b) 0.30%

(c) 0.12%

5-11. **(a)** $(-0.4/30) \times 100\% = -1.3\%$

As in part (a)

(c) -0.13%

5-12. $\text{mean} = \left(\dfrac{0.0110 + 0.0104 + 0.0105}{3} \right) = 0.01063 \approx 0.0106$

Arranging the numbers in increasing value the median is:

  0.0104

  0.0105 $\leftarrow$ median

  0.0110

The deviations from the mean are:

$$|0.0104 - 0.01063| = 0.00023$$
$$|0.0105 - 0.01063| = 0.00013$$
$$|0.0110 - 0.01063| = 0.00037$$

$\text{mean deviation} = \left(\dfrac{0.00023 + 0.00013 + 0.00037}{3} \right) = 0.00024 \approx 0.0002$

(c) mean = 190   median = 189   mean deviation = 2

  deviations 1.75, 0.25, 4.25, 2.75. rounded to 2, 0, 4, 3

(e) mean = 39.59   median = 39.65   mean deviation = 0.17

  rounded deviations 0.24, 0.02, 0.34, 0.09

Chapter 6

6-1.   **(a)** The *standard error of the mean* is the standard deviation of the mean and is given by

the standard deviation of the data set divided by the square root of the number of

measurements.

(c) The *variance* is the square of the standard deviation.

6-2.   **(a)** The term *parameter* refers to quantities such as the mean and standard deviation of a

population or distribution of data. The term *statistic* refers to an estimate of a parameter

that is made from a sample of data.

(c) *Random errors* result from uncontrolled variables in an experiment while *systematic*

errors are those that can be ascribed to a particular cause and can usually be determined.

6-3.   **(a)** The *sample standard deviation s* is the standard deviation of a sample drawn from the

population. It is given by $s = \sqrt{\dfrac{\sum\limits_{i=1}^{N}(x_i - \bar{x})^2}{N-1}}$, where $\bar{x}$ is the sample mean.

The *population standard deviation* σ is the standard deviation of an entire population

given by $\sigma = \sqrt{\dfrac{\sum\limits_{i=1}^{N}(x_i - \mu)^2}{N}}$, where μ is the population mean.

6-5.   Since the probability that a result lies between -1σ and +1σ is 0.683, the probability that a

result will lie between 0 and +1σ will be half this value or 0.342. The probability that a

result will lie between +1σ and +2σ will be half the difference between the probability of

the result being between -2σ and +2σ, and -1σ and +1σ, or ½ (0.954-0.683) = 0.136.

6-7. Listing the data from Set A in order of increasing value:

x_i	x_i^2
9.5	90.25
8.5	72.25
9.1	82.81
9.3	86.49
9.1	82.81
$\Sigma x_i = 45.5$	$\Sigma x_i^2 = 414.61$

(a) mean: $\bar{x} = 45.5/5 = 9.1$

(b) median = 9.1

(c) spread: $w = 9.5 - 8.5 = 1.0$

(d) standard deviation: $s = \sqrt{\dfrac{414.61 - (45.5)^2/5}{5-1}} = 0.37$

(e) coefficient of variation: $CV = (0.37/9.1) \times 100\% = 4.1\%$

Results for Sets A through F, obtained in a similar way, are given in the following table.

	A	B	C	D	E	F
$\bar{x}$	9.1	55.29	0.650	5.1	20.61	0.958
median	9.1	55.32	0.653	5.0	20.64	0.954
w	1.0	0.15	0.108	1.5	0.14	0.049
s	0.37	0.08	0.056	0.6	0.07	0.02
CV, %	4.1	0.14	8.5	12.2	0.32	2.1

6-8. For Set A, $\quad E = 9.1 - 9.0 = 0.1$

$E_r = (0.1/9.0) \times 1000 \text{ ppt} = 11.1 \text{ ppt}$

Set C $\quad E = 0.0195 \qquad\qquad E_r = 31 \text{ ppt}$

Set E $\quad E = 0.03 \qquad\qquad\quad E_r = 1.3 \text{ ppt}$

6-9. **(a)** $s_y = \sqrt{(0.03)^2 + (0.001)^2 + (0.001)^2} = 0.030$

$CV = (0.03/{-}2.082) \times 100\% = -1.4\%$

$y = -2.08(\pm 0.03)$

16

(c) $\dfrac{s_y}{y} = \sqrt{\left(\dfrac{0.3}{29.2}\right)^2 + \left(\dfrac{0.02\times10^{-17}}{2.034\times10^{-17}}\right)^2} = 0.01422$

$CV = (0.0142) \times 100\% = 1.42\%$

$s_y = (0.0142) \times (5.93928 \times 10^{-16}) = 0.08446 \times 10^{-16}$

$y = 5.94(\pm0.08) \times 10^{-16}$

(e) $s_{num} = \sqrt{(6)^2 + (3)^2} = 6.71$ $y_{num} = 187 - 89 = 98$

$s_{den} = \sqrt{(1)^2 + (8)^2} = 8.06$ $y_{den} = 1240 + 57 = 1297$

$\dfrac{s_y}{y} = \sqrt{\left(\dfrac{6.71}{98}\right)^2 + \left(\dfrac{8.06}{1297}\right)^2} = 0.0688$

$CV = (0.0688) \times 100\% = 6.88\%$

$s_y = (0.0688) \times (0.075559) = 0.00520$

$y = 7.6(\pm0.5) \times 10^{-2}$

6-10. (a) $s_y = \sqrt{(0.02\times10^{-8})^2 + (0.2\times10^{-9})^2} = 2.83\times10^{-10}$

$y = 1.02\times10^{-8} - 3.54\times10^{-9} = 6.66\times10^{-9}$

$CV = \dfrac{2.83\times10^{-10}}{6.66\times10^{-9}} \times 100\% = 4.25\%$

$y = 6.7 \pm0.3 \times 10^{-9}$

(c) $\dfrac{s_y}{y} = \sqrt{\left(\dfrac{0.0005}{0.0040}\right)^2 + \left(\dfrac{0.02}{10.28}\right)^2 + \left(\dfrac{1}{347}\right)} = 0.1250$

$CV = (0.1250) \times 100\% = 12.5\%$

$y = 0.0040 \times 10.28 \times 347 = 14.27$

$$s_y = (0.125) \times (14.27) = 1.78$$

$$y = 14(\pm 2)$$

(e) $\quad \dfrac{s_y}{y} = \sqrt{\left(\dfrac{1}{100}\right)^2 + \left(\dfrac{1}{2}\right)^2} = 0.500$

$$CV = (0.500) \times 100\% = 50.0\%$$

$$y = 100 \, / \, 2 = 50.0$$

$$s_y = (0.500) \times (50.0) = 25$$

$$y = 50(\pm 25)$$

6-11. **(a)** $\quad y = \log(2.00 \times 10^{-4}) = -3.6989 \qquad s_y = \dfrac{(0.434)(0.03 \times 10^{-4})}{(2.00 \times 10^{-4})} = 6.51 \times 10^{-3}$

$$y = -3.699 \pm 0.0065$$

$$CV = (0.0065/3.699) \times 100\% = 0.18\%$$

(c) $\quad y = \text{antilog}(1.200) = 15.849 \qquad \dfrac{s_y}{y} = (2.303)(0.003) = 0.0069$

$$s_y = (0.0069)(15.849) = 0.11 \qquad y = 15.8 \pm 0.1$$

$$CV = (0.11/15.8) \times 100\% = 0.69\%$$

6-12.   (a) $y = (4.17 \times 10^{-4})^3 = 7.251 \times 10^{-11}$ $\dfrac{s_y}{y} = 3\left(\dfrac{0.03 \times 10^{-4}}{4.17 \times 10^{-4}}\right) = 0.0216$

$s_y = (0.0216)(7.251 \times 10^{-11}) = 1.565 \times 10^{-12}$ $y = 7.3(\pm 0.2) \times 10^{-11}$

$CV = (1.565 \times 10^{-12}/7.251 \times 10^{-11}) \times 100\% = 2.2\%$

6-13.   From the equation for the volume of a sphere, we have

$$V = \frac{4}{3}\pi r^3 = \frac{4}{3}\pi\left(\frac{d}{2}\right)^3 = \frac{4}{3}\pi\left(\frac{2.15}{2}\right)^3 = 5.20 \text{ cm}^3$$

Hence, we may write

$$\frac{s_V}{V} = 3 \times \frac{s_d}{d} = 3 \times \frac{0.02}{2.15} = 0.0279$$

$$s_V = 5.20 \times 0.0279 = 0.145$$

$$V = 5.2(\pm 0.1) \text{ cm}^3$$

6-15.   Since the titrant volume equals the final buret reading minus the initial buret reading, we

can introduce the values given into the equation for %A.

$\%A = [9.26(\pm 0.03) - 0.19(\pm 0.02)] \times$ equivalent mass $\times 100/[45.0(\pm 0.2)]$

Obtaining the value of the first term and the error in the first term

$s_y = \sqrt{(0.03)^2 + (0.02)^2} = 0.0361$ $y = 9.26 - 0.19 = 9.07$

We can now obtain the relative error of the calculation

$$\frac{s_{\%A}}{\%A} = \sqrt{\left(\frac{0.036}{9.07}\right)^2 + \left(\frac{0.2}{45.0}\right)^2} = 0.00596$$

The coefficient of variation is then

$CV = (0.00596) \times 100\% = 0.596\%$ or 0.6%

6-17. We first calculate the mean transmittance and the standard deviation of the mean.

$$\text{mean T} = \left(\frac{0.213+0.216+0.208+0.214}{4}\right) = 0.2128$$

$s_T = 0.0034$

(a) $c_X = \left(\dfrac{-\log T}{\varepsilon b}\right) = \dfrac{-\log(0.2128)}{3312} = 2.029 \times 10^{-4}\text{ M}$

(b) For $-\log T$, $s_y = (0.434)s_T/T = 0.434 \times (0.0034/0.2128) = 0.00693$

$-\log(0.2128) = 0.672$

$$c_X = \frac{-\log T}{\varepsilon b} = \frac{0.672 \pm 0.00693}{3312 \pm 12}$$

$$\frac{s_{C_X}}{c_X} = \sqrt{\left(\frac{0.00693}{0.672}\right)^2 + \left(\frac{12}{3312}\right)^2} = 0.0109$$

$$s_{C_X} = (0.0109)(2.029 \times 10^{-4}) = 2.22 \times 10^{-6}$$

(c) CV $= (2.22 \times 10^{-6}/2.029 \times 10^{-4}) \times 100\% = 1.1\%$

6-19.

	A	B	C	D	E	F	G	H	I	J	K	L	M
1	Problem 6-19												
2													
3	Sample	1	$(x_i-x_{ave})^2$	2	$(x_i-x_{ave})^2$	3	$(x_i-x_{ave})^2$	4	$(x_i-x_{ave})^2$	5	$(x_i-x_{ave})^2$	6	$(x_i-x_{ave})^2$
4													
5		1.02	0.0049	1.13	0.0020	1.12	0.0071	0.77	0.0100	0.73	0.0144	0.73	0.0008
6		0.84	0.0121	1.02	0.0042	1.32	0.0135	0.58	0.0081	0.92	0.0049	0.88	0.0150
7		0.99	0.0016	1.17	0.0072	1.13	0.0055	0.61	0.0036	0.90	0.0025	0.72	0.0014
8				1.02	0.0042	1.20	0.0000	0.72	0.0025			0.70	0.0033
9						1.25	0.0021						
10													
11	mean	0.950		1.085		1.204		0.670		0.850		0.758	
12	s	0.096		0.077		0.084		0.090		0.104		0.083	
13	N		3		4		5		4		3		4
14	$\Sigma(x_i-x_{ave})^2$		0.0186		0.0177		0.0281		0.0242		0.0218		0.0205
15													
16	s_{pooled}	0.088								No. Sets	6		
17													
18	Spreadsheet Documentation									N_{Total}	23		
19										$\Sigma(x_i-x_{ave})^2$	0.1309		
20	B11=AVERAGE(B5:B9)												
21	B12=STDEV(B5:B9)												
22	C5=(B5-$B$11)^2												
23	C13=COUNT(C5:C9)												
24	C14=SUM(C5:C9)												
25	K18=SUM(C13:M13)												
26	K19=SUM(C14:M14)												
27	B16=SQRT(K19/(K18-K16))												

(a) The standard deviations are $s_1 = 0.096$, $s_2 = 0.077$, $s_3 = 0.084$, $s_4 = 0.090$, $s_5 = 0.104$, $s_6 = 0.083$

(b) $s_{pooled} = 0.088$ or 0.09

6-21.

	A	B	C	D	E	F
1	Problem 6-20					
2						
3	Sample	x_1	x_2	mean	$(x_1-x_{ave})^2$	$(x_2-x_{ave})^2$
4	1	2.24	2.27	2.255	0.00022	0.00023
5	2	8.4	8.7	8.55	0.02250	0.02250
6	3	7.6	7.5	7.55	0.00250	0.00250
7	4	11.9	12.6	12.25	0.12250	0.12250
8	5	4.3	4.2	4.25	0.00250	0.00250
9	6	1.07	1.02	1.045	0.00063	0.00062
10	7	14.4	14.8	14.6	0.04000	0.04000
11	8	21.9	21.1	21.5	0.16000	0.16000
12	9	8.8	8.4	8.6	0.04000	0.04000
13						
14	N	18		Total	0.39085	0.39085
15	No. of Sets	9				
16	s_{pooled}	0.29				
17						
18	Spreadsheet Documentation					
19						
20	D4=AVERAGE(B4:C4)					
21	E4=(B4-$D$4)^2					
22	F4=(C4-$D4)^2					
23	B14=COUNT(B4:C12)					
24	E14=SUM(E4:E12)					
25	B16=SQRT((E14+F14)/(B14-B15))					

Chapter 7

7-1. The distribution of means is narrower than the distribution of single results. Hence, the standard error of the mean of 5 measurements is smaller than the standard deviation of a single result. The mean is thus known with more confidence than is a single result.

7-4. For Set A

x_i	x_i^2
2.7	7.29
3.0	9.00
2.6	6.76
2.8	7.84
3.2	10.24
$\Sigma x_i = 14.3$	$\Sigma x_i^2 = 41.13$

mean: $\bar{x} = 14.3/5 = 2.86$

standard deviation: $s = \sqrt{\dfrac{41.13 - (14.3)^2/5}{5-1}} = 0.24$

Since, for a small set of measurements we cannot be certain s is a good approximation of σ, we should use the t statistic for confidence intervals. From Table 7-3, at 95% confidence t for 4 degrees of freedom is 2.78, therefore for set A,

$$\text{CI for } \mu = 2.86 \pm \frac{(2.78)(0.24)}{\sqrt{5}} = 2.86 \pm 0.30$$

Similarly, for the other data sets, we obtain the results shown in the following table:

	A	C	E
$\bar{x}$	2.86	70.19	0.824
s	0.24	0.08	0.051
CI	2.86 ±0.30	70.19 ±0.20	0.824 ±0.081

The 95% confidence interval is the range within which the population mean is expected

to lie with a 95% probability.

7-5.   If s is a good estimate of σ then we can use $z = 1.96$ for the 95% confidence level. For

set A, at the 95% confidence,

$$\text{CI for } \mu = 2.86 \pm \frac{(1.96)(0.30)}{\sqrt{5}} = 2.86 \pm 0.26. \text{ Similarly for sets C and E, the limits are:}$$

	A	C	E
CI	2.86±0.26	70.19±0.079	0.824±0.009

7-7.   **(a)** 99% CI = $18.5 \pm 2.58 \times 3.6 = 18.5 \pm 9.3$ µg Fe/mL

   95% CI = $18.5 \pm 1.96 \times 3.6 = 18.5 \pm 7.1$ µg Fe/mL

   (b) 99% CI = $18.5 \pm \dfrac{2.58 \times 3.6}{\sqrt{2}} = 18.5 \pm 6.6$ µg Fe/mL

   95% CI = $18.5 \pm \dfrac{1.96 \times 3.6}{\sqrt{2}} = 18.5 \pm 5.0$ µg Fe/mL

   (c) 99% CI = $18.5 \pm \dfrac{2.58 \times 3.6}{\sqrt{4}} = 18.5 \pm 4.6$ µg Fe/mL

   95% CI = $18.5 \pm \dfrac{1.96 \times 3.6}{\sqrt{4}} = 18.5 \pm 3.5$ µg Fe/mL

7-9.　$2.2 = \dfrac{1.96 \times 3.6}{\sqrt{N}}$　　For a 95% CI, $N = 10.3 \cong 11$

$2.2 = \dfrac{2.58 \times 3.6}{\sqrt{N}}$　　For a 99% CI, $N = 17.8 \cong 18$

7-11. For the data set, $\bar{x} = 3.22$ and $s = 0.06$

(a) 95% CI $= 3.22 \pm \dfrac{4.30 \times 0.06}{\sqrt{3}} = 3.22 \pm 0.15$ meq Ca/L

(b) 95% CI $= 3.22 \pm \dfrac{1.96 \times 0.056}{\sqrt{3}} = 3.22 \pm 0.06$ meq Ca/L

7-13. **(a)** $0.3 = \dfrac{2.58 \times 0.38}{\sqrt{N}}$ For the 99% CI, $N = 10.7 \cong 11$

7-15. This is a two-tailed test where $s \rightarrow \sigma$ and from Table 7-1, $z_{crit} = 2.58$ for the 99% confidence level.

For As:　　$z = \dfrac{129 - 119}{9.5\sqrt{\dfrac{3+3}{3 \times 3}}} = 1.28 \le 2.58$

No significant difference exists at the 99% confidence level.

Proceeding in a similar fashion for the other elements

Element	z	Significant Difference?
As	1.28	No
Co	−3.43	Yes
La	2.45	No
Sb	0.20	No
Th	-3.42	Yes

For two of the elements there is a significant difference, but for three there are not. Thus, the defendant might have grounds for claiming reasonable doubt. It would be prudent, however, to analyze other windows and show that these elements are good diagnostics for

the rare window.

7-17. $Q = \dfrac{|5.6-5.1|}{5.6-4.3} = 0.385$ and Q_{crit} for 8 observations at 95% confidence = 0.526.

Since $Q < Q_{crit}$ the outlier value 5.6 cannot be rejected at the 95% confidence level.

7-19.   The null hypothesis is that for the pollutant the current level = the previous level (H_0:

$\mu_{current} = \mu_{previous}$). The alternative hypothesis is H_a: $\mu_{current} > \mu_{previous}$ This would be a

one-tailed test. The type I error for this situation would be that we reject the null

hypothesis when, in fact, it is true, i.e. we decide the level of the pollutant is > the

previous level at some level of confidence when, in fact, it is not. The type II error would

be that we accept the null hypothesis when, in fact, it is false, i.e. we decide the level of

the pollutant = the previous level when, in fact, it is > than the previous level.

7-20.   (a) H_0: $\mu_{ISE} = \mu_{EDTA}$, H_a: $\mu_{ISE} \neq \mu_{EDTA}$. This would be a two-tailed test. The type I error

for this situation would be that we decide the methods agree when they do not. The type

II error would be that we decide the methods do not agree when they do.

(c) H_0: $\sigma_X^2 = \sigma_Y^2$; H_a $\sigma_X^2 < \sigma_Y^2$. This is a one-tailed test. The type I error would be that

we decide that $\sigma_X^2 < \sigma_Y^2$ when it is not. The type II error would be that we decide that

$\sigma_X^2 = \sigma_Y^2$ when actually $\sigma_X^2 < \sigma_Y^2$.

7-21.   (a) For the Top data set, $\bar{x} = 26.338$

For the bottom data set, $\bar{x} = 26.254$

$s_{pooled} = 0.1199$

degrees of freedom $= 5 + 5 - 2 = 8$

26

For 8 degrees of freedom at 95% confidence $t_{crit} = 2.31$

$$t = \frac{26.338 - 26.254}{0.1199\sqrt{\frac{5+5}{5\times5}}} = 1.11$$ Since $t < t_{crit}$, we conclude that no significant difference

exists at 95% confidence level.

(b) From the data, $N = 5$, $\bar{d} = 0.084$ and $s_d = 0.015166$

For 4 degrees of freedom at 95% confidence $t = 2.78$

$$t = \frac{0.084 - 0}{0.015/\sqrt{5}} = 12.52$$

Since $12.52 > 2.78$, a significant difference does exist at 95% confidence level.

(c) The large sample to sample variability causes s_{Top} and s_{Bottom} to be large and masks

the differences between the samples taken from the top and the bottom.

7-23. For the first data set: $\bar{x} = 2.2978$

For the second data set: $\bar{x} = 2.3106$

$s_{pooled} = 0.0027$

Degrees of freedom $= 4 + 3 - 2 = 5$

$$t = \frac{2.2978 - 2.3106}{0.0027\sqrt{\frac{4+3}{4\times3}}} = -6.207$$

For 5 degrees of freedom at the 99% confidence level, $t = 4.03$ and at the 99.9%

confidence level, $t = 6.87$. Thus, we can be between 99% and 99.9% confident that the

nitrogen prepared in the two ways is different. The Excel TDIST(x,df,tails) function can

be used to calculate the probability of getting a t value of -6.207. In this case we find

TDIST(6.207,5,2) = 0.0016. Therefore, we can be 99.84% confident that the nitrogen prepared in the two ways is different. There is a 0.16% probability of this conclusion

being in error.

7-25 (a)

Source	SS	df	MS	F
Between juices	$4 \times 7.715 = 30.86$	$5 - 1 = 4$	$0.913 \times 8.45 = 7.715$	8.45
Within juices	$25 \times 0.913 = 22.825$	$30 - 5 = 25$	0.913	
Total	$30.86 + 22.82 = 50.68$	$30 - 1 = 29$		

(b) H_0: $\mu_{brand1} = \mu_{brand2} = \mu_{brand3} = \mu_{brand4} = \mu_{brand5}$; H_a: at least two of the means differ.

(c) The Excel FINV(prob,df1,df2) function can be used to calculate the F value for the

above problem. In this case we find FINV(0.05,4,25) = 2.76. Since F calculated exceeds

F critical, we reject the null hypothesis and conclude that the average ascorbic acid

contents of the 5 brands of orange juice differ at the 95% confidence level.

7-27.

(a) H_0: $\mu_{Analyst1} = \mu_{Analyst2} = \mu_{Analyst3} = \mu_{Analyst4}$; H_a: at least two of the means differ.

(b) See spreadsheet next page. From Table 7-4 the F value for 3 degrees of freedom in

the numerator and 12 degrees of freedom in the denominator at 95% is 3.49. Since F

calculated exceeds F critical, we reject the null hypothesis and conclude that the analysts

differ at 95% confidence. The F value calculated of 13.60 also exceeds the critical values

at the 99% and 99.9% confidence levels so that we can be certain that the analysts differ

at these confidence levels.

(c) Based on the calculated LSD value there is a significant difference between analyst 2

and analysts 1 and 4, but not analyst 3. There is a significant difference between analyst

3 and analyst 1, but not analyst 4. There is a significant difference between analyst 1 and

analyst 4.

Spreadsheet for Problem 7-27.

	A	B	C	D	E	F	G
1	Detmn	Analys 1	Analyst 2	Analyst 3	Analyst 4		
2	1	10.24	10.14	10.19	10.19		
3	2	10.26	10.12	10.11	10.15		
4	3	10.29	10.04	10.15	10.16		
5	4	10.23	10.07	10.12	10.10		
6							
7	Mean	10.26	10.09	10.14	10.15		
8	Std. Dev.	0.02646	0.04573	0.03594	0.03742		
9	Variance	0.00070	0.00209	0.00129	0.00140		
10							
11	Grand Mean	10.16					
12	SSF	0.05595		**Differences**			
13	SSE	0.01645		10.26-10.09=	0.17	Significant difference	
14	SST	0.07240		10.15-10.09=	0.06	Significant difference	
15				10.14-10.09=	0.05	No sig. diff.	
16	MSF	0.01865		10.26-10.14=	0.12	Significant difference	
17	MSE	0.001371		10.15-10.14=	0.01	No sig. diff.	
18				10.26-10.15=	0.11	Significant difference	
19	F	13.60486					
20							
21	LSD	0.057335					
22							
23	**Spreadsheet Documentation**						
24	B7=AVERAGE(B2:B5)						
25	B8=STDEV(B2:B5)						
26	B9=VAR(B2:B5)						
27	B11=AVERAGE(B2:E5)						
28	B12=4*((B7-B11)^2+(C7-B11)^2+(D7-B11)^2+(E7-B11)^2)						
29	B13=3*SUM(B9:E9)						
30	B14=B12+B13						
31	B16=B12/3						
32	B17=B13/12						
33	B19=B16/B17						
34	B21=2.19*SQRT(2*B17/4)						

7-29.　(a) H_0: $\mu_{ISE} = \mu_{EDTA} = \mu_{AA}$; H_a: at least two of the means differ.

(b) See Spreadsheet

	A	B	C	D	E	F	G
1	**Repetition**	**ISE**	**EDTA**	**At. Abs.**			
2	1	39.2	29.9	44.0			
3	2	32.8	28.7	49.2			
4	3	41.8	21.7	35.1			
5	4	35.3	34.0	39.7			
6	5	33.5	39.1	45.9			
7							
8	Mean	36.52	30.68	42.78			
9	Std. Dev.	3.85707	6.46313	5.49791			
10	Variance	14.877	41.772	30.227			
11							
12	Grand Mean	36.660		**Differences**			
13	SSF	366.172		42.78-30.68=		12.1 Significant difference	
14	SSE	347.504		36.52-30.68=		5.94 No sig. diff.	
15	SST	713.676		42.78-36.52=		6.26 No sig. diff.	
16							
17	MSF	183.086					
18	MSE	28.95867					
19	F	6.322321					
20	LSD	7.453554					
21							
22	**Spreadsheet Documentation**						
23	B8=AVERAGE(B2:B6)						
24	B9=STDEV(B2:B6)						
25	B10=VAR(B2:B6)						
26	B12=AVERAGE(B2:D6)						
27	B13=5*((B8-B12)^2+(C8-B12)^2+(D8-B12)^2)						
28	B14=4*SUM(B10:D10)						
29	B15=B13+B14						
30	B17=B13/2						
31	B18=B14/12						
32	B19=B17/B18						
33	B20=2.19*SQRT(2*B18/5)						

From Table 7-4 the F value for 2 degrees of freedom in the numerator and 12 degrees of freedom in the denominator at 95% is 3.89. Since F calculated is greater than F critical, we reject the null hypothesis and conclude that the 3 methods give different results at the 95% confidence level.

(c) Based on the calculated LSD value there is a significant difference between the atomic absorption method and the EDTA titration. There is no significant difference between the EDTA titration method and the ion-selective electrode method and there is no significant difference between the atomic absorption method and the ion-selective electrode method.

7-31. **(a)** $Q = \dfrac{|85.10 - 84.70|}{85.10 - 84.62} = 0.833$ and Q_{crit} for 3 observations at 95% confidence = 0.970.

Since $Q < Q_{crit}$ the outlier value 85.10 cannot be rejected with 95% confidence.

(b) $Q = \dfrac{|85.10 - 84.70|}{85.10 - 84.62} = 0.833$ and Q_{crit} for 4 observations at 95% confidence = 0.829.

Since $Q > Q_{crit}$ the outlier value 85.10 can be rejected with 95% confidence.

Chapter 8

8-1.   The sample size is in the micro range and the analyte level is in the trace range. Hence, the analysis is a micro analysis of a trace constituent.

8-3.   Step 1: Identify the population from which the sample is to be drawn.

Step 2: Collect the gross sample.

Step 3: Reduce the gross sample to a laboratory sample, which is a small quantity of homogeneous material

8-5. $s_o^2 = s_s^2 + s_m^2$

From the NIST sample: $s_m^2 = 0.00947$

From the gross sample: $s_o^2 = 0.15547$

$s_s = \sqrt{0.15547 - 0.00947} = 0.38$

The relative standard deviation $= \left(\dfrac{s_s}{\overline{x}}\right) \times 100\% = \left(\dfrac{0.38}{49.92}\right) \times 100\% = 0.76\%$

8-7.   **(a)** $N = \dfrac{(1-p)}{p\sigma_r^2} = \dfrac{(1-0.02)}{0.02(0.20)^2} = \dfrac{49.0}{(0.20)^2} = 1225$

(b) $N = 49.0/(0.12)^2 = 3403$

(c) $N = 49.0/(0.07)^2 = 10000$

(d) $N = 49.0/(0.02)^2 = 122500$

32

8-9. $\quad N = p(1-p)\left(\dfrac{d_A d_B}{d^2}\right)^2\left(\dfrac{P_A - P_B}{\sigma_r P}\right)^2$

    (a) $d = 7.3 \times 0.15 + 2.6 \times 0.85 = 3.3$

    $P = 0.15 \times 7.3 \times 0.87 \times 100 / 3.3 = 29\%$

    $N = 0.15(1-0.15)\left(\dfrac{7.3 \times 2.6}{(3.3)^2}\right)^2\left(\dfrac{87-0}{0.020 \times 29}\right)^2 = 8714$ particles

    (b) mass $= (4/3)\pi(r)^3 \times d \times N = (4/3)\pi(0.175 \text{ cm})^3 \times 3.3(\text{g/cm}^3) \times 8.714 \times 10^3$

        $= 650$ g

    (c) $0.500 = (4/3)\pi(r)^3 \times 3.3(\text{g/cm}^3) \times 8.714 \times 10^3$

        $r = 0.016$ cm   (diameter $= 0.32$ mm)

8-11.   (a) The following single-factor ANOVA table was generated using Excel's Data

Analysis Tools:

Anova: Single Factor

SUMMARY

Groups	Count	Sum	Average	Variance
1	3	185	61.66667	2.333333
2	3	172	57.33333	0.333333
3	3	146	48.66667	4.333333
4	3	170	56.66667	6.333333

ANOVA

Source of Variation	SS	df	MS	F	P-value	F crit
Between Groups	264.25	3	88.08333	26.425	0.000167	4.066181
Within Groups	26.66667	8	3.333333			
Total	290.9167	11				

The Between Groups *SS* value of 264.25 compared to the Within Groups value of

26.66667 indicates that the mean concentrations vary significantly from day to day.

(b)   SST is the total variance and is the sum of the within day variance, SSE, and the day-to-

day variance, SSF; SST = SSE + SSF. The within day variance, SSE, reflects the method

variance, SSM. The day-to-day variance, SSF, reflects the sum of the method variance,

SSM, and the sampling variance, SSS; SSF = SSM + SSS. Thus,

SST = SSM + SSM + SSS   and   SSS = SST – 2×SSM

SSS = 290.92 – 2×26.67 = 237.58. Dividing 3 degrees of freedom gives a mean square

(estimates sampling variance σ_s^2) of 79.19.

(c)   The best approach to lowering the overall variance would be to reduce the sampling

variance, since this is the major component of the total variance (σ_t^2 = 88.08333).

8-13.   See Example 8-3

Using $t = 1.96$ for infinite samples $N = \dfrac{(1.96)^2 \times (0.3)^2}{(3.7)^2 \times (0.07)^2} = 5.16$

Using $t = 2.78$ for 5 samples (4 df) $N = \dfrac{(2.78)^2 \times (0.3)^2}{(3.7)^2 \times (0.07)^2} = 10.36$

Using $t = 2.26$ for 10 samples $N = \dfrac{(2.26)^2 \times (0.3)^2}{(3.7)^2 \times (0.07)^2} = 6.85$

Using $t = 2.45$ for 7 samples $N = \dfrac{(2.45)^2 \times (0.3)^2}{(3.7)^2 \times (0.07)^2} = 8.05$

Using $t = 2.36$ for 8 samples $N = \dfrac{(2.36)^2 \times (0.3)^2}{(3.7)^2 \times (0.07)^2} = 7.47$

The iterations converge at between 7 and 8 samples, so 8 should be taken for safety.

8-15.

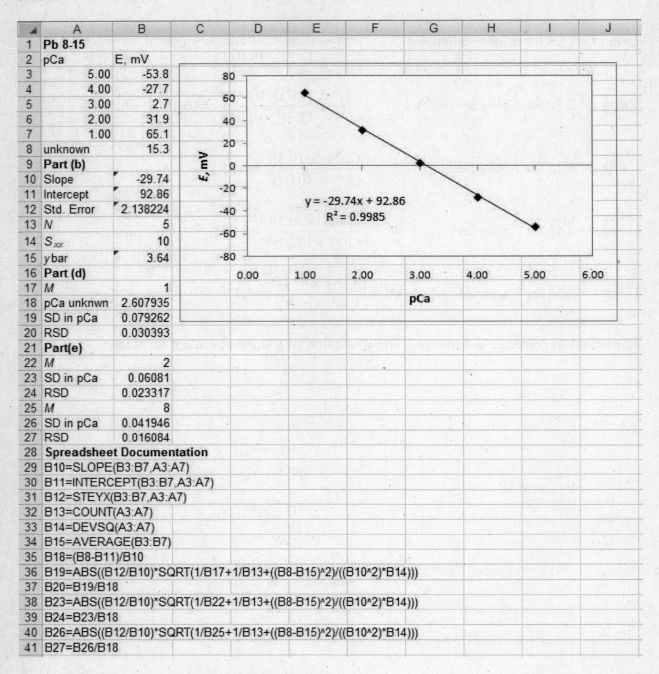

	A	B
1	**Pb 8-15**	
2	pCa	E, mV
3	5.00	-53.8
4	4.00	-27.7
5	3.00	2.7
6	2.00	31.9
7	1.00	65.1
8	unknown	15.3
9	**Part (b)**	
10	Slope	-29.74
11	Intercept	92.86
12	Std. Error	2.138224
13	N	5
14	S_{xx}	10
15	ybar	3.64
16	**Part (d)**	
17	M	1
18	pCa unknwn	2.607935
19	SD in pCa	0.079262
20	RSD	0.030393
21	**Part(e)**	
22	M	2
23	SD in pCa	0.06081
24	RSD	0.023317
25	M	8
26	SD in pCa	0.041946
27	RSD	0.016084
28	**Spreadsheet Documentation**	
29	B10=SLOPE(B3:B7,A3:A7)	
30	B11=INTERCEPT(B3:B7,A3:A7)	
31	B12=STEYX(B3:B7,A3:A7)	
32	B13=COUNT(A3:A7)	
33	B14=DEVSQ(A3:A7)	
34	B15=AVERAGE(B3:B7)	
35	B18=(B8-B11)/B10	
36	B19=ABS((B12/B10)*SQRT(1/B17+1/B13+((B8-B15)^2)/((B10^2)*B14)))	
37	B20=B19/B18	
38	B23=ABS((B12/B10)*SQRT(1/B22+1/B13+((B8-B15)^2)/((B10^2)*B14)))	
39	B24=B23/B18	
40	B26=ABS((B12/B10)*SQRT(1/B25+1/B13+((B8-B15)^2)/((B10^2)*B14)))	
41	B27=B26/B18	

(b) Equation of the line: $y = -29.74 \, x + 92.8$

(d) $pCa_{Unk} = 2.608$; SD in pCa $= 0.079$; RSD $= 0.030$ (CV $= 3.0\%$)

8-17.

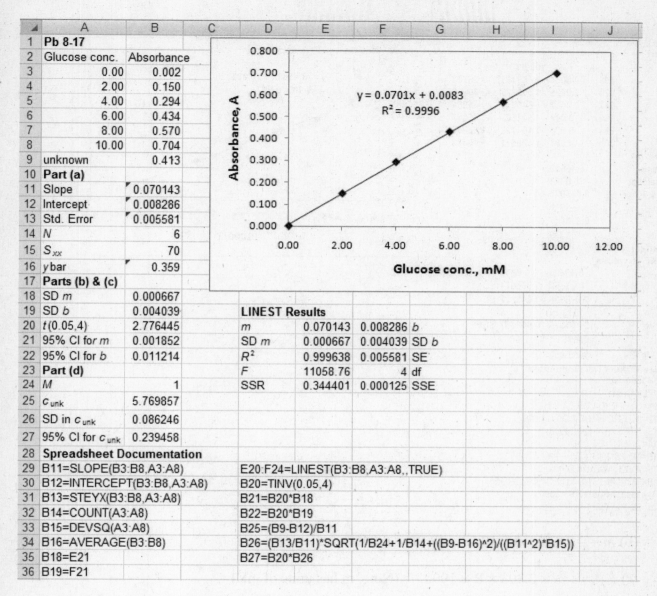

	A	B	C	D	E	F	G	H	I	J
1	**Pb 8-17**									
2	Glucose conc.	Absorbance								
3	0.00	0.002								
4	2.00	0.150								
5	4.00	0.294								
6	6.00	0.434								
7	8.00	0.570								
8	10.00	0.704								
9	unknown	0.413								
10	**Part (a)**									
11	Slope	0.070143								
12	Intercept	0.008286								
13	Std. Error	0.005581								
14	N	6								
15	S_{xx}	70								
16	y bar	0.359								
17	**Parts (b) & (c)**									
18	SD m	0.000667								
19	SD b	0.004039		**LINEST Results**						
20	t(0.05,4)	2.776445		m	0.070143	0.008286	b			
21	95% CI for m	0.001852		SD m	0.000667	0.004039	SD b			
22	95% CI for b	0.011214		R^2	0.999638	0.005581	SE			
23	**Part (d)**			F	11058.76	4	df			
24	M	1		SSR	0.344401	0.000125	SSE			
25	c_{unk}	5.769857								
26	SD in c_{unk}	0.086246								
27	95% CI for c_{unk}	0.239458								
28	**Spreadsheet Documentation**									
29	B11=SLOPE(B3:B8,A3:A8)			E20:F24=LINEST(B3:B8,A3:A8,,TRUE)						
30	B12=INTERCEPT(B3:B8,A3:A8)			B20=TINV(0.05,4)						
31	B13=STEYX(B3:B8,A3:A8)			B21=B20*B18						
32	B14=COUNT(A3:A8)			B22=B20*B19						
33	B15=DEVSQ(A3:A8)			B25=(B9-B12)/B11						
34	B16=AVERAGE(B3:B8)			B26=(B13/B11)*SQRT(1/B24+1/B14+((B9-B16)^2)/((B11^2)*B15))						
35	B18=E21			B27=B20*B26						
36	B19=F21									

(a)　　$m = 0.07014$ and $b = 0.008286$

(b)　　$s_m = 0.00067$; $s_b = 0.004039$; SE $= 0.00558$

(c)　　95% $CI_m = m \pm t \times s_m = 0.07014 \pm 0.0019$

　　　　95% $CI_b = b \pm t \times s_b = 0.0083 \pm 0.0112$

(d)　　$c_{unk} = 5.77$ mM; $s_{unk} = 0.09$; 95% $CI_{unk} = c_{unk} \pm t \times s_{Unk} = 5.77 \pm 0.24$ mM

8-19.

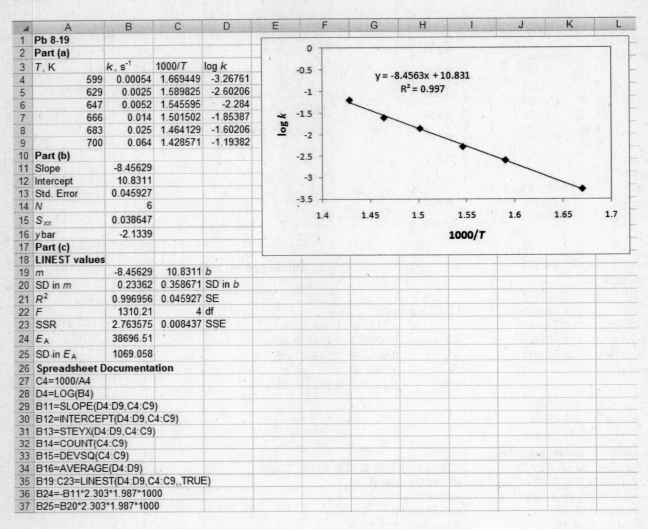

	A	B	C	D	E	F	G	H	I	J	K	L
1	Pb 8-19											
2	Part (a)											
3	T, K	k, s^{-1}	1000/T	log k								
4	599	0.00054	1.669449	-3.26761								
5	629	0.0025	1.589825	-2.60206								
6	647	0.0052	1.545595	-2.284								
7	666	0.014	1.501502	-1.85387								
8	683	0.025	1.464129	-1.60206								
9	700	0.064	1.428571	-1.19382								
10	Part (b)											
11	Slope	-8.45629										
12	Intercept	10.8311										
13	Std. Error	0.045927										
14	N	6										
15	S_{xx}	0.038647										
16	y bar	-2.1339										
17	Part (c)											
18	LINEST values											
19	m	-8.45629	10.8311	b								
20	SD in m	0.23362	0.358671	SD in b								
21	R^2	0.996956	0.045927	SE								
22	F	1310.21	4	df								
23	SSR	2.763575	0.008437	SSE								
24	E_A	38696.51										
25	SD in E_A	1069.058										
26	Spreadsheet Documentation											
27	C4=1000/A4											
28	D4=LOG(B4)											
29	B11=SLOPE(D4:D9,C4:C9)											
30	B12=INTERCEPT(D4:D9,C4:C9)											
31	B13=STEYX(D4:D9,C4:C9)											
32	B14=COUNT(C4:C9)											
33	B15=DEVSQ(C4:C9)											
34	B16=AVERAGE(D4:D9)											
35	B19:C23=LINEST(D4:D9,C4:C9,,TRUE)											
36	B24=-B11*2.303*1.987*1000											
37	B25=B20*2.303*1.987*1000											

(b) $m = -8.456$; $b = 10.83$ and SE = 0.0459

(c) $E_A = -m \times 2.303 \times R \times 1000$ (Note: m has units of mK) =

$-(-8.456 \text{ mK}) \times (2.303) \times (1.987 \text{ cal mol}^{-1} \text{ K}^{-1}) \times (1000 \text{ K/mK})$

$= 38697$ cal/mol

$s_{EA} = s_m \times 2.303 \times R \times 1000$

$= 1069$ cal/mol

Thus, $E_A = 38,697 \pm 1069$ cal/mol or 38.7 ± 1.1 kcal/mol

(d) H_0: E_A = 41.00 kcal/mol; H_A: $E_A \neq 41.00$ kcal/mol.

$$t = (38.697 - 41.00)/1.069 = -2.15$$

$$t(0.025, 4) = 2.776$$

Since $t > -t_{crit}$ we retain H_0. There is no reason to doubt that E_A is not 41.00 kcal/mol at

the 95% confidence level.

8-21. **(c)** 5.247 ppm rounded to 5.2 ppm

8-23. See Example 8-8

$$c_u = \frac{(0.300)(1.00 \times 10^{-3})(1.00)}{(0.530)(51.00) - (0.300)(50.00)} = 2.4938 \times 10^{-5} \text{ M}$$

To obtain the concentration of the original sample, we need to multiply by 25.00/1.00.

$$c_u = (2.4938 \times 10^{-5} \text{ M})(25.00)/(1.00) = 6.23 \times 10^{-4} \text{ M}$$

8-25.

(c)   For $k = 2$, DL = 0.14 ng/mL (92.1% confidence level)

for $k = 3$, DL = 0.21 ng/mL (98.3% confidence level)

8-27.

	A	B	C	D	E	F	G	H	I	J	K	L	M	N
1	Pb 8-27													
2	Day	Mean	SD	$\Sigma(x_i-x_{ave})^2$	UCL	LCL								
3	1	96.50	0.80	3.20	98.08	94.97								
4	2	97.38	0.88	3.87	98.08	94.97								
5	3	96.85	1.43	10.22	98.08	94.97								
6	4	96.64	1.59	12.64	98.08	94.97								
7	5	96.87	1.52	11.55	98.08	94.97								
8	6	95.52	1.27	8.06	98.08	94.97								
9	7	96.08	1.16	6.73	98.08	94.97								
10	8	96.48	0.79	3.12	98.08	94.97								
11	9	96.63	1.48	10.95	98.08	94.97								
12	10	95.47	1.30	8.45	98.08	94.97								
13	11	97.38	0.88	3.87	98.08	94.97								
14	12	96.85	1.43	10.22	98.08	94.97								
15	13	96.64	1.59	12.64	98.08	94.97								
16	14	96.87	1.52	11.55	98.08	94.97								
17	15	95.52	1.27	8.06	98.08	94.97								
18	16	96.08	1.16	6.73	98.08	94.97								
19	17	96.48	0.79	3.12	98.08	94.97								
20	18	96.63	1.48	10.95	98.08	94.97								
21	19	95.47	1.30	8.45	98.08	94.97								
22	20	96.43	0.75	2.81	98.08	94.97								
23	21	97.06	1.34	8.98	98.08	94.97								
24	22	98.34	1.60	12.80	98.08	94.97								
25	23	96.42	1.22	7.44	98.08	94.97								
26	24	95.99	1.18	6.96	98.08	94.97								
27														
28	Mean	96.52		Spreadsheet Documentation										
29	N	24		B28=AVERAGE(B3:B26)			D3=C3^2*5							
30	s_{pooled}	1.269523		B29=COUNT(A3:A26)			E3=$B$31							
31	UCL	98.07901		B30=SQRT(SUM(D3:D26)/(B29*6-B29))			F3=$B$32							
32	LCL	94.96933		B31=B28+3*B30/SQRT(6)										
33				B32=B28-3*B30/SQRT(6)										

The process went out of control on Day 22.

Chapter 9

9-1.   **(a)** A *weak electrolyte* only partially ionizes when dissolved in water. H_2CO_3 is an

example of a weak electrolyte.

(c) The *conjugate acid of a Brønsted-Lowry base* is the potential proton donor formed

when a Brønsted-Lowry base accepts a proton. For example, the NH_4^+ is a conjugate

acid in the reaction, $NH_3 + proton \rightleftharpoons NH_4^+$.

(e) An *amphiprotic solvent* can act either as an acid or a base depending on the solute.

Water is an example of an amphiprotic chemical species.

(g) *Autoprotolysis* is the act of self-ionization to produce both a conjugate acid and a

conjugate base.

(i) The *Le Châtelier principle* states that the position of an equilibrium always shifts in

such a direction that it relieves the stress. A common ion like sulfate added to a solution

containing sparingly soluble $BaSO_4$ is an example

9-2.   **(a)** An *amphiprotic solute* is a chemical species that possesses both acidic and basic

properties. The dihydrogen phosphate ion, $H_2PO_4^-$, is an example of an amphiprotic

solute.

(c) A *leveling solvent* shows no difference between strong acids. Perchloric acid and

hydrochloric acid ionize completely in water; thus, water is a leveling solvent.

9-3.   For dilute aqueous solutions, the concentration of water remains constant and is assumed

to be independent of the equilibrium. Thus, its concentration is included within the

equilibrium constant. For a pure solid, the concentration of the chemical species in the

solid phase is constant. As long as some solid exists as a second phase, its effect on the

equilibrium is constant and is included within the equilibrium constant.

9-4.

	Acid	Conjugate Base
(a)	HOCl	OCl$^-$
(c)	NH$_4^+$	NH$_3$
(e)	H$_2$PO$_4^-$	HPO$_4^-$

9-6. **(a)** $2H_2O \rightleftharpoons H_3O^+ + OH^-$

(c) $2CH_3NH_2 \rightleftharpoons CH_3NH_3^+ + CH_3NH^-$

9-7. **(a)** $C_2H_5NH_2 + H_2O \rightleftharpoons C_2H_5NH_2^+ + OH^-$

$$K_b = \frac{K_w}{K_a} = \frac{1.00 \times 10^{-14}}{2.3 \times 10^{-11}} = \frac{[C_2H_5NH_2^+][OH^-]}{[C_2H_5NH_2]} = 4.3 \times 10^{-4}$$

(c)

$$CH_3NH_3^+ + H_2O \rightleftharpoons CH_3NH_2 + H_3O^+$$

$$K_a = \frac{K_w}{K_b} = \frac{[CH_3NH_2][H_3O^+]}{[CH_3NH_3^+]} = 2.3 \times 10^{-11}$$

(e) $H_3AsO_4 + H_2O \rightleftharpoons H_3O^+ + H_2AsO_4^-$

$H_2AsO_4^- + H_2O \rightleftharpoons H_3O^+ + HAsO_4^{2-}$

$\underline{HAsO_4^{2-} + H_2O \rightleftharpoons H_3O^+ + AsO_4^{3-}}$

$H_3AsO_4 + 3H_2O \rightleftharpoons 3H_3O^+ + AsO_4^{3-}$

$$K_{a1} = \frac{[H_3O^+][H_2AsO_4^-]}{[H_3AsO_4]} \quad K_{a2} = \frac{[H_3O^+][HAsO_4^{2-}]}{[H_2AsO_4^-]} \quad K_{a3} = \frac{[H_3O^+][AsO_4^{3-}]}{[HAsO_4^{2-}]}$$

$$K_{overall} = \frac{[H_3O^+]^3[AsO_4^{3-}]}{[H_3AsO_4]} = K_{a1}K_{a2}K_{a3} = 5.8 \times 10^{-3} \times 1.1 \times 10^{-7} \times 3.2 \times 10^{-12} = 2.0 \times 10^{-21}$$

solubility product expression (handwritten)

9-8. (a) $CuBr(s) \rightleftharpoons Cu^+ + Br^-$ $K_{sp} = [Cu^+][Br^-]$

(b) $HgCII(s) \rightleftharpoons Hg^{2+} + Cl^- + I^-$ $K_{sp} = [Hg^{2+}][Cl^-][I^-]$

(c) $PbCl_2(s) \rightleftharpoons Pb^{2+} + 2Cl^-$ $K_{sp} = [Pb^{2+}][Cl^-]^2$

9-10. (b) $RaSO_4 \rightleftharpoons Ra^{2+} + SO_4^{2-}$ *Calc. solubility product constant* (handwritten)

$[Ra^{2+}] = [SO_4^{2-}] = 6.6 \times 10^{-6}$ M

$K_{sp} = [Ra^{2+}][SO_4^{2-}] = (6.6 \times 10^{-6} \text{ M})^2 = 4.4 \times 10^{-11}$

(d) $Ce(IO_3)_3 \rightleftharpoons Ce^{3+} + 3IO_3^-$

$[Ce^{3+}] = 1.9 \times 10^{-3}$ M $[IO_3^-] = 3 \times 1.9 \times 10^{-3}$ M $= 5.7 \times 10^{-3}$ M

$K_{sp} = [Ce^{3+}][IO_3^-]^3 = 1.9 \times 10^{-3} \times (5.7 \times 10^{-3})^3 = 3.5 \times 10^{-10}$

9-13. $Ag_2CrO_4(s) \rightleftharpoons 2Ag^+ + CrO_4^{2-}$

(a) $[CrO_4^{2-}] = \dfrac{1.2 \times 10^{-12}}{(4.13 \times 10^{-3})^2} = 7.04 \times 10^{-8}$ M *Ag* (handwritten)

(b) $[CrO_4^{2-}] = \dfrac{1.2 \times 10^{-12}}{(9.00 \times 10^{-7})^2} = 1.48$ M *Ag* (handwritten)

9-15. $Ce^{3+} + 3IO_3^- \rightleftharpoons Ce(IO_3)_3(s)$ *What is the Ce^{3+}?* (handwritten)

$K_{sp} = [Ce^{3+}][IO_3^-]^3 = 3.2 \times 10^{-10}$

(a) 50.00 mL $\times 0.0450$ mmol/ml $= 2.25$ mmol Ce^{3+}

$[Ce^{3+}] = \dfrac{2.25 \text{ mmol}}{(50.00 + 50.00) \text{ mL}} = 0.0225$ M

(b) We mix 2.25 mmol Ce^{3+} with 50.00 mL $\times$ 0.045 mmol/mL = 2.25 mmol IO_3^-.

Each mole of IO_3^- reacts with 1/3 mole of Ce^{3+} so 2.25 mmol would consume $1/3 \times 2.25$

mmol Ce^{3+} or the amount of unreacted Ce^{3+} = 2.25 − 2.25/3 = 1.5 mmol

$$c_{Ce^{3+}} = \frac{1.50 \text{ mmol}}{100 \text{ mL}} = 0.0150 \text{ M}$$

$[Ce^{3+}] = 0.0150 + S$　　　(where S is the solubility).

$[IO_3^-] = 3S$

$K_{sp} = [Ce^{3+}][IO_3^-]^3 = 0.0150 \times 3S^3 = 3.2 \times 10^{-10}$

$$S = \left(\frac{3.2 \times 10^{-10}}{27 \times 1.50 \times 10^{-2}} \right)^{1/3} = 9.2 \times 10^{-4}$$

$[Ce^{3+}] = 1.50 \times 10^{-2} + 9.2 \times 10^{-4} = 1.6 \times 10^{-2} \text{ M}$

(c) Now we have 0.250 mmol $IO_3^- \times 50.00$ mL = 12.5 mmol. Since 3×2.25 mmol =

6.75 mmol would be required to completely react with the Ce^{3+}, we have excess IO_3^-.

$$[IO_3^-] = \frac{12.5 \text{ mmol} - 6.75 \text{ mmol}}{100 \text{ mL}} + 3S = 0.0575 + 3S$$

$[Ce^{3+}] = S$

$K_{sp} = S(0.0575 + 3S)^3 = 3.2 \times 10^{-10}$

Lets assume $3S << 0.0575$

$S = [Ce^{3+}] = 3.2 \times 10^{-10}/(0.0575)^3 = 1.7 \times 10^{-6} \text{ M}$

Checking the assumption $3 \times 1.7 \times 10^{-6} \text{ M} = 5.1 \times 10^{-6}$ which is much smaller than

0.0575.

(d) Now we are mixing 2.25 mmol Ce^{3+} with $50.00 \text{ mL} \times 0.050 \text{ mmol/mL} = 2.50$ mmol

IO_3^-. The Ce^{3+} is now in excess so that

amount of Ce^{3+} = 2.25 mmol − 2.5/3 mmol = 1.42 mmol

$$c_{Ce^{3+}} = \frac{1.42 \text{ mmol}}{100 \text{ mL}} = 0.0142 \text{ M}$$

$[Ce^{3+}] = 1.42 \times 10^{-2} + S$

$K_{sp} = [Ce^{3+}][IO_3^-]^3 = 0.0142 \times 3S^3 = 3.2 \times 10^{-10}$

45

$$S = \left(\frac{3.2 \times 10^{-10}}{27 \times 1.42 \times 10^{-2}} \right)^{1/3} = 9.42 \times 10^{-4}$$

$$[Ce^{3+}] = 1.42 \times 10^{-2} + 9.42 \times 10^{-4} = 1.5 \times 10^{-2} \text{ M}$$

9-17. $CuI(s) \rightleftharpoons Cu^+ + I^-$ $K_{sp} = [Cu^+][I^-] = 1 \times 10^{-12}$

$AgI(s) \rightleftharpoons Ag^+ + I^-$ $K_{sp} = [Ag^+][I^-] = 8.3 \times 10^{-17}$

$PbI_2(s) \rightleftharpoons Pb^{2+} + 2I^-$ $K_{sp} = [Pb^{2+}][I^-]^2 = 7.1 \times 10^{-9} = S(2S)^2 = 4S^3$

$BiI_3(s) \rightleftharpoons Bi^{3+} + 3I^-$ $K_{sp} = [Bi^{3+}][I^-]^3 = 8.1 \times 10^{-19} = S(3S)^3 = 27S^4$

(a) For CuI, $S = [Cu^+] = [I^-] = \sqrt{1 \times 10^{-12}} = 1 \times 10^{-6} \text{ M}$

For AgI, $S = [Ag^+] = [I^-] = \sqrt{8.3 \times 10^{-17}} = 9.1 \times 10^{-9} \text{ M}$

For PbI$_2$, $S = \sqrt[3]{\dfrac{7.1 \times 10^{-9}}{4}} = 1.2 \times 10^{-3} \text{ M}$

For BiI$_3$ $S = \sqrt[4]{\dfrac{8.1 \times 10^{-19}}{27}} = 1.3 \times 10^{-5} \text{ M}$

So, solubilities are in the order PbI$_2$ > BiI$_3$ > CuI > AgI

(b) For CuI, $S = 1 \times 10^{-12}/0.20 = 5 \times 10^{-12} \text{ M}$

For AgI, $S = 8.3 \times 10^{-17}/0.20 = 4.2 \times 10^{-16} \text{ M}$

For PbI$_2$, $S = 7.1 \times 10^{-9}/(0.20)^2 = 1.8 \times 10^{-7} \text{ M}$

For BiI$_3$, $S = 8.1 \times 10^{-19}/(0.20)^3 = 1.0 \times 10^{-16} \text{ M}$

So, solubilities are in the order PbI$_2$ > CuI > AgI > BiI$_3$

(c) For CuI, $S = 1 \times 10^{-12}/0.020 = 5 \times 10^{-11} \text{ M}$

For AgI, $S = 8.3 \times 10^{-17}/0.020 = 4.2 \times 10^{-15} \text{ M}$

For PbI$_2$, $S = \dfrac{1}{2}\sqrt{\dfrac{7.1 \times 10^{-9}}{0.020}} = 3.0 \times 10^{-4}$ M

For BiI$_3$, $S = \dfrac{1}{3}\sqrt[3]{\dfrac{8.1 \times 10^{-19}}{0.020}} = 1.1 \times 10^{-6}$ M

So solubilities are in the order, PbI$_2$ > BiI$_3$ > CuI > AgI

9-19. At 25°C, pK_w = 13.99, K_w = 1.023 × 10^{-14}. At 75°C, pK_w = 12.70, K_w = 1.995 × 10^{-13}

[H$_3$O$^+$] = [OH$^-$] in pure water. Thus [H$_3$O$^+$] = $\sqrt{K_w}$

At 25°C, [H$_3$O$^+$] = $\sqrt{1.023 \times 10^{-14}}$ = 1.011 × 10^{-7} M, pH = $-\log$[H$_3$O$^+$] = 6.99$_5$ ≈ 7.00

At 75°C, [H$_3$O$^+$] = $\sqrt{1.995 \times 10^{-13}}$ = 4.467 × 10^{-7} M, pH = 6.35

9-20. **(a)** For benzoic acid, K_a = 6.28 × 10^{-5}. Call benzoic acid HBz and the benzoate anion

Bz$^-$

$$HBz + H_2O \rightleftharpoons Bz^- + H_3O^+$$

$K_a = \dfrac{[Bz^-][H_3O^+]}{[HBz]} = 6.28 \times 10^{-5}$ Mass balance c_{HBz} = [HBz] + [Bz$^-$] = 0.0300

[Bz$^-$] = [H$_3$O$^+$] Thus, [HBz] = 0.0300 − [Bz$^-$] = 0.0300 − [H$_3$O$^+$]

$$\dfrac{[H_3O^+]^2}{0.0300 - [H_3O^+]} = 6.28 \times 10^{-5}$$

Solving the quadratic or solving by iterations gives,

[H$_3$O$^+$] = 1.34 × 10^{-3} M so [OH$^-$] = 1.00 × 10^{-14}/ 1.34 × 10^{-3} = 7.5 × 10^{-12} M

(c)

$$C_2H_5NH_2 + H_2O \rightleftharpoons C_2H_5NH_3^+ + OH^-$$

$$K_b = \frac{[C_2H_5NH_3^+][OH^-]}{[C_2H_5NH_2]} = \frac{K_w}{K_a} = \frac{1.0\times10^{-14}}{2.31\times10^{-11}} = 4.33\times10^{-4}$$

$$[OH^-] = [C_2H_5NH_3^+] \qquad [C_2H_5NH_2] = 0.100 - [OH^-]$$

$$\frac{[OH^-]^2}{(0.100 - [OH^-])} = 4.33\times10^{-4} \qquad [OH^-]^2 = 4.33\times10^{-4}(0.100 - [OH^-])$$

$$[OH^-]^2 + 4.33\times10^{-4}[OH^-] - 4.33\times10^{-5} = 0$$

$$[OH^-] = -\frac{4.33\times10^{-4} + \sqrt{(4.33\times10^{-4})^2 + 4(4.33\times10^{-5})}}{2} = 6.37\times10^{-3}\,M$$

$$[H_3O^+] = \frac{1.0\times10^{-14}}{6.37\times10^{-3}} = 1.57\times10^{-12}\,M$$

(e)

$$Bz^- + H_2O \rightleftharpoons HBz + OH^- \qquad K_b = K_w/K_a = 1.00\times10^{-14}/6.28\times10^{-5} = 1.60\times10^{-10}$$

$$[OH^-] = [HBz] \qquad [Bz^-] = 0.200 - [OH^-]$$

$$\frac{[OH^-]^2}{0.200 - [OH^-]} = 1.60\times10^{-10}$$

$$[OH^-] = 5.66\times10^{-6}\,M \qquad [H_3O^+] = 1.00\times10^{-14}/5.66\times10^{-6} = 1.77\times10^{-9}\,M$$

(g) $HONH_3^+ + H_2O \rightleftharpoons HONH_2 + H_3O^+ \qquad K_a = 1.1\times10^{-6}$

As in part (b) $\dfrac{[H_3O^+]^2}{0.250 - [H_3O^+]} = 1.1\times10^{-6}$

$$[H_3O^+] = 5.24\times10^{-4}\,M \qquad [OH^-] = 1.91\times10^{-11}\,M$$

9-21. (a)

$$ClCH_2COOH + H_2O \rightleftharpoons ClCH_2COO^- + H_3O^+ \qquad K_a = \frac{[ClCH_2COO^-][H_3O^+]}{[ClCH_2COOH]} = 1.36 \times 10^{-3}$$

$$[H_3O^+] = [ClCH_2COO^-] \qquad [ClCH_2COOH] = 0.200 - [H_3O^+]$$

$$\frac{[H_3O^+]^2}{(0.200 - [H_3O^+])} = 1.36 \times 10^{-3} \quad [H_3O^+]^2 = 1.36 \times 10^{-3}(0.200 - [H_3O^+])$$

$$[H_3O^+]^2 + 1.36 \times 10^{-3}[H_3O^+] - 2.72 \times 10^{-4} = 0$$

$$[H_3O^+] = -\frac{1.36 \times 10^{-3} + \sqrt{(1.36 \times 10^{-3})^2 + 4(2.72 \times 10^{-4})}}{2} = 1.58 \times 10^{-2} \text{ M}$$

(b)

$$ClCH_2COO^- + H_2O \rightleftharpoons ClCH_2COOH + OH^-$$

$$K_b = \frac{[ClCH_2COOH][OH^-]}{[ClCH_2COO^-]} = \frac{K_w}{K_a} = \frac{1.0 \times 10^{-14}}{1.36 \times 10^{-3}} = 7.35 \times 10^{-12}$$

$$[OH^-] = [ClCH_2COOH] \qquad [ClCH_2COO^-] = 0.200 \, M - [OH^-]$$

$$\frac{[OH^-]^2}{(0.200 - [OH^-])} = 7.35 \times 10^{-12} \quad [OH^-]^2 = 7.35 \times 10^{-12}(0.200 - [OH^-])$$

$$[OH^-]^2 + 7.35 \times 10^{-12}[OH^-] - 1.47 \times 10^{-12} = 0$$

$$[OH^-] = 1.21 \times 10^{-6} \text{ M}$$

$$[H_3O^+] = \frac{1.0 \times 10^{-14}}{1.21 \times 10^{-6}} = 8.26 \times 10^{-9} \text{ M}$$

(e)

$$C_6H_5NH_3^+ + H_2O \rightleftharpoons C_6H_5NH_2 + H_3O^+ \qquad K_a = \frac{[C_6H_5NH_2][H_3O^+]}{[C_6H_5NH_3^+]} = 2.51 \times 10^{-5}$$

$$[H_3O^+] = [C_6H_5NH_2] \qquad\qquad [C_6H_5NH_3^+] = 0.0020 \, M - [H_3O^+]$$

$$\frac{[H_3O^+]^2}{(0.0020 - [H_3O^+])} = 2.51 \times 10^{-5}$$

Proceeding as in part (d), we find $[H_3O^+] = 2.11 \times 10^{-4}$ M

9-23. *Buffer capacity* of a solution is defined as the number of moles of a strong acid (or a

strong base) that causes 1.00 L of a buffer to undergo a 1.00-unit change in pH.

49

9-25. $HOAc + H_2O \rightleftharpoons OAc^- + H_3O^+$ $OAc^- + H_2O \rightleftharpoons HOAc + OH^-$

(a)

$$[OAc^-] = \frac{8.00 \text{ mmol}}{200 \text{ mL}} = 4 \times 10^{-2} \text{ M}$$

$$[HOAc] = 0.100 \text{ M}$$

$$pH = -\log(1.75 \times 10^{-5}) + \log\frac{4 \times 10^{-2}}{0.100} = 4.359$$

(b)

$$0.175 \text{ M HOAc} = \frac{0.175 \text{ mmol}}{\text{mL}} \times 100 \text{ mL} = 17.5 \text{ mmol}$$

$$0.0500 \text{ M NaOH} = \frac{0.0500 \text{ mmol}}{\text{mL}} \times 100 \text{ mL} = 5.00 \text{ mmol}$$

$$[HOAc] = \frac{(17.5 - 5.00) \text{ mmol}}{200 \text{ mL}} = 6.25 \times 10^{-2} \text{ M}$$

$$[OAc^-] = \frac{5 \text{ mmol}}{200 \text{ mL}} = 2.50 \times 10^{-2} \text{ M}$$

$$pH = -\log(1.75 \times 10^{-5}) + \log\frac{2.50 \times 10^{-2}}{6.25 \times 10^{-2}} = 4.359$$

(c)

$$0.0420 \text{ M OAc}^- = \frac{0.042 \text{ mmol}}{\text{mL}} \times 160 \text{ mL} = 6.72 \text{ mmol}$$

$$0.1200 \text{ M HCl} = \frac{0.1200 \text{ mmol}}{\text{mL}} \times 40.0 \text{ mL} = 4.80 \text{ mmol}$$

$$[OAc^-] = \frac{(6.72 - 4.80) \text{ mmol}}{200 \text{ mL}} = 9.6 \times 10^{-3} \text{ M}$$

$$[HOAc] = \frac{4.8 \text{ mmol}}{200 \text{ mL}} = 2.4 \times 10^{-2} \text{ M}$$

$$pH = -\log(1.75 \times 10^{-5}) + \log\frac{9.6 \times 10^{-3}}{2.4 \times 10^{-2}} = 4.359$$

The solutions all are buffers with the same pH, but they differ in buffer capacity with **(a)** having the greatest and **(c)** the least.

9-26. **(a)** $C_6H_5NH_3^+/C_6H_5NH_2$ ($pK_a = 4.60$)

(c) The closest are $C_2H_5NH_3^+/ C_2H_5NH_2$ ($pK_a = 10.64$) and $CH_3NH_3^+/CH_3NH_2$ $pK_a = 10.64$)

9-27.

$$pH = 3.50 = pK_a + \log \frac{[HCOO^-]}{[HCOOH]} = -\log(1.8 \times 10^{-4}) + \log \frac{[HCOO^-]}{[HCOOH]}$$

$$3.50 = 3.74 + \log \frac{[HCOO^-]}{[HCOOH]} \qquad \frac{[HCOO^-]}{[HCOOH]} = 10^{-0.24} = 0.575$$

$$500 \text{ mL} \times 1.00 \frac{\text{mmol HCOOH}}{\text{mL}} = 500 \text{ mmol}$$

So amount of $HCOO^-$ needed = 0.575×500 mmol = 287.5 mmol

287.5 mmol $\times 10^{-3}$ mol/mmol = 0.2875 mol $HCOO^-$

Mass HCOONa = 0.2875 mol $\times$ 67.997 g/mol = 19.6 g

9-29. Let HMn = mandelic acid, Mn^- = mandelate anion.

500 mL $\times$ 0.300 M NaMn = 150 mmol Mn^-.

For a pH of 3.37 need the ratio of Mn^- to HMn to be

$$pH = 3.37 = pK_a + \log \frac{[Mn^-]}{[HMn]} = 3.398 + \log \frac{[Mn^-]}{[HMn]} \qquad \log \frac{[Mn^-]}{[HMn]} = 3.37 - 3.398 = -0.028$$

$$\frac{[Mn^-]}{[HMn]} = 0.938$$

$$\frac{\text{mmol } Mn^- - x \text{ mmol HCl}}{x \text{ mmol HCl}} = 0.938$$

$0.938 \times x$ mmol HCl = mmol $Mn^- - x$ mmol HCl

x = mmol $Mn^-/1.938$ = 150 $Mn^-/1.938$ = 77.399 mmol HCl

Volume HCl = 77.399 mmol/(0.200 mmol/mL) = 387 mL

Chapter 10 Make a distinction between

10-1. **(a)** *Activity*, a_A, is the effective concentration of a chemical species A in solution. The

activity coefficient, γ_A, is the numerical factor necessary to convert the molar

concentration of the chemical species A to activity as shown below:

$a_A = \gamma_A[A]$

(b) The *thermodynamic equilibrium constant* refers to an ideal system within which each

chemical species is unaffected by any others. A *concentration equilibrium constant* takes

into account the influence exerted by solute species upon one another. The

thermodynamic equilibrium constant is numerically constant and independent of ionic

strength; the concentration equilibrium constant depends on molar concentrations of

reactants and products as well as other chemical species that may not participate in the

equilibrium.

10-3. **(a)** $MgCl_2 + 2NaOH \rightleftharpoons Mg(OH)_2(s) + 2NaCl$

Replacing divalent Mg^{2+} with Na^+ causes the ionic strength to decrease.

(b) $HCl + NaOH \rightleftharpoons H_2O + NaCl$

There is no change in the charge states of the ions present in the solution equilibria. The

ionic strength is unchanged.

(c) $HOAc + NaOH \rightleftharpoons H_2O + NaOAc$

The ionic strength will increase because NaOH and NaOAc are totally ionized wheras

acetic acid is only partially ionized.

10-5. Water is a neutral molecule and its activity equals its concentration at all low to moderate

ionic strengths. That is, its activity coefficient is unity. In solutions of low to moderate ionic

strength, activity coefficients of ions decrease with increasing ionic strength because the ionic atmosphere surrounding the ion causes it to lose some of its chemical effectiveness and its

activity is less than its concentration.

10-7. Multiply charged ions deviate from ideality more than singly charged ions because of the

effect of the surrounding ionic atmosphere. The initial slope of the activity coefficient vs

square root of ionic strength for Ca^{2+} is steeper than that for K^+ the activity coefficient of

Ca^{2+} is more influenced by ionic strength than that for K^+.

10-9. **(a)** $\mu = \frac{1}{2}[0.030 \times 2^2 + 0.030 \times 2^2] = 0.12$ *Calc. Ionic Strength μ*

   (c) $\mu = \frac{1}{2}[0.30 \times 3^2 + 0.90 \times 1^2 + 0.20 \times 2^2 + 0.40 \times 1^2] = 2.4$

10-10. $-\log \gamma_x = \dfrac{0.51 Z_x^2 \sqrt{\mu}}{1 + 3.3\alpha_x \sqrt{\mu}}$ This problem is easiest to work with a spreadsheet.

Calc. Activity Coefficient

	A	B	C	D	E	F
1	**Problem 10-10**					
2	Ion X	Fe^{3+}	Pb^{2+}	Ce^{4+}	Sn^{4+}	
3	Z	3	2	4	4	
4	μ	0.062	0.042	0.07	0.045	
5	α_X	0.9	0.45	1.1	1.1	
6	$\log \gamma_X$	0.657	0.3205	1.1013	0.9779	
7	γ_X	0.2203	0.478	0.0792	0.1052	
8						
9	**Documentation**					
10	Cell B6=0.51*B3^2*SQRT(B4)/(1+3.3*B5*SQRT(B4))					
11	Cell B7=10^-B6					

Rounding these results gives

(a) 0.22   **(c)** 0.08

10-12. We must use $-\log \gamma_X = \dfrac{0.51 Z_X^2 \sqrt{\mu}}{1 + 3.3 \alpha_X \sqrt{\mu}}$

(a) For Ag^+, $\alpha_{Ag^+} = 0.25$. At $\mu = 0.08$, $\gamma_{Ag^+} = 0.7639$; For SCN^-, $\alpha_{SCN^-} = 0.35$ and γ_{SCN^-}

$= 0.7785$ retaining insignificant figures for later calculations.

$$K'_{sp} = \frac{K_{sp}}{\gamma_{Ag^+}\gamma_{SCN^-}} = \frac{1.1 \times 10^{-12}}{(0.7639)(0.7785)} = 1.8 \times 10^{-12}$$

(c) For La^{3+}, $\gamma_{La^{3+}} = 0.197$. For IO_3^-, $\gamma_{IO3^-} = 0.7785$

$$K'_{sp} = \frac{K_{sp}}{\gamma_{La^{3+}}\gamma_{IO_3^-}} = \frac{1.0 \times 10^{-11}}{(0.197)(0.7785)^3} = 1.1 \times 10^{-10}$$

10-13. $Zn(OH)_2(s) \rightleftharpoons Zn^{2+} + 2OH^-$ $K_{sp} = 3.0 \times 10^{-16}$

(a) $\mu = \frac{1}{2}[0.02 \times 1^2 + 0.02 \times 1^2] = 0.02$

Using Equation 10-5,

$\gamma_{Zn^{2+}} = 0.5951$ $\gamma_{OH^-} = 0.867$

$K'_{sp} = a_{Zn^{2+}} a_{OH^-}^2 = \gamma_{Zn^{2+}}[Zn^{2+}] \times \gamma_{OH^-}^2 [OH^-]^2$

$[Zn^{2+}][OH^-]^2 = \dfrac{3.0 \times 10^{-16}}{\gamma_{Zn^{2+}} \gamma_{OH^-}^2} = \dfrac{3.0 \times 10^{-16}}{(0.5951)(0.867)^2} = 6.706 \times 10^{-16}$

Solubility $= S = [Zn^{2+}] = \frac{1}{2}[OH^-]$

$S(2S)^2 = 6.706 \times 10^{-16}$

$S = \left(\dfrac{6.706 \times 10^{-16}}{4} \right)^{1/3} = 5.5 \times 10^{-6} \text{ M}$

(b) $\mu = \frac{1}{2}[2 \times 0.03 \times 1^2 + 0.03 \times 2^2] = 0.18$

From Equation 10-5,

$$\gamma_{Zn^{2+}} = 0.3386 \quad \gamma_{OH^-} = 0.7158$$

$$K'_{sp} = a_{Zn^{2+}} a^2_{OH^-} = \gamma_{Zn^{2+}}[Zn^{2+}] \times \gamma^2_{OH^-}[OH^-]^2$$

$$[Zn^{2+}][OH^-]^2 = \frac{3.0 \times 10^{-16}}{(0.3386)(0.7158)^2} = 1.729 \times 10^{-15}$$

Solubility $= S = [Zn^{2+}] = \frac{1}{2}[OH^-]$

$$S(2S)^2 = 1.729 \times 10^{-15}$$

$$S = \left(\frac{1.729 \times 10^{-15}}{4}\right)^{\frac{1}{3}} = 7.6 \times 10^{-6} \text{ M}$$

(c)

$$\text{amount of KOH} = \frac{0.250 \text{ mmol}}{\text{mL}} \times 40.0 \text{ mL} = 10.0 \text{ mmol}$$

$$\text{amount of ZnCl}_2 = \frac{0.0250 \text{ mmol}}{\text{mL}} \times 60.0 \text{ mL} = 1.5 \text{ mmol}$$

$$[K^+] = \frac{10 \text{ mmol}}{100.0 \text{ mL}} = 0.10 \text{ M}$$

$$[OH^-] = \frac{(10 \text{ mmol} - (2 \times 1.5 \text{ mmol}))}{100.0 \text{ mL}} \times = 0.07 \text{ M}$$

$$[Cl^-] = \frac{2 \times 1.5 \text{ mmol}}{100.0 \text{ mL}} = 0.03 \text{ M}$$

$[Zn^{2+}] = 0$

$$\mu = \frac{1}{2}[0.10 \times 1^2 + 0.07 \times 1^2 + 2 \times 0.03 \times 1^2] = 0.115$$

From Equation 10-5,

$$\gamma_{Zn^{2+}} = 0.3856 \quad \gamma_{OH^-} = 0.7511 \quad K'_{sp} = a_{Zn^{2+}} a^2_{OH^-} = \gamma_{Zn^{2+}}[Zn^{2+}] \times \gamma^2_{OH^-}[OH^-]^2$$

$$[Zn^{2+}][OH^-]^2 = \frac{3.0 \times 10^{-16}}{\gamma_{Zn^{2+}} \gamma^2_{OH^-}} = \frac{3.0 \times 10^{-16}}{(0.3856)(0.7511)^2} = 1.379 \times 10^{-15}$$

Solubility $= S = [Zn^{2+}] \quad S(0.07)^2 = 1.379 \times 10^{-15}$

$$S = \left(\frac{1.379 \times 10^{-15}}{(0.07)^2}\right) = 2.8 \times 10^{-13} \text{ M}$$

(d)

$$\text{amount KOH} = \frac{0.100 \text{ mmol}}{\text{mL}} \times 20.0 \text{ mL} = 2.0 \text{ mmol}$$

$$\text{amount ZnCl}_2 = \frac{0.0250 \text{ mmol}}{\text{mL}} \times 80.0 \text{ mL} = 2.0 \text{ mmol}$$

$$[K^+] = \frac{2 \text{ mmol}}{100.0 \text{ mL}} = 0.02 \text{ M}$$

$$[OH^-] = 0$$

$$[Cl^-] = \frac{2 \times 2.0 \text{ mmol}}{100.0 \text{ mL}} = 0.04 \text{ M}$$

$$[Zn^{2+}] = \frac{2 \text{ mmol} - \frac{1}{2}(2 \text{ mmol})}{100.0 \text{ mL}} = 0.01 \text{ M}$$

$$\mu = \frac{1}{2}\left(0.02 \times 1^2 + 0.040 \times 1^2 + 0.01 \times 2^2\right) = 0.05$$

From Table 10-2,

$$\gamma_{Zn^{2+}} = 0.48 \quad \gamma_{OH^-} = 0.81$$

$$K'_{sp} = a_{Zn^{2+}} a^2_{OH^-} = \gamma_{Zn^{2+}}[Zn^{2+}] \times \gamma^2_{OH^-}[OH^-]^2$$

$$[Zn^{2+}][OH^-]^2 = \frac{3.0 \times 10^{-16}}{\gamma_{Zn^{2+}}\gamma^2_{OH^-}} = \frac{3.0 \times 10^{-16}}{(0.48)(0.81)^2} = 9.53 \times 10^{-16}$$

Solubility $= S = [OH^-]/2$

$$(0.01)[OH^-]^2 = 9.53 \times 10^{-16}$$

$$[OH^-] = \left(\frac{9.53 \times 10^{-16}}{0.01}\right)^{\frac{1}{2}} = 3.09 \times 10^{-7} \text{ M}$$

$$S = \left(3.09 \times 10^{-7} \text{ M}\right)/2 = 1.5 \times 10^{-7} \text{ M}$$

10-14. $\mu = \frac{1}{2}[0.0333 \times 2^2 + 2 \times 0.0333 \times 1^2] = 0.100$ Can use data in Table 10-2.

(a) $AgSCN(s) \rightleftharpoons Ag^+ + SCN^-$

(1) For Ag^+, $\gamma_{Ag^+} = 0.75$; for SCN^-, $\gamma_{SCN^-} = 0.76$

$$K'_{sp} = \gamma_{Ag^+}[Ag^+]\gamma_{SCN^-}[SCN^-] = 1.1 \times 10^{-12}$$

$$[Ag^+][SCN^-] = \frac{1.1 \times 10^{-12}}{0.75 \times 0.76} = 1.9298 \times 10^{-12}$$

$$S = [Ag^+] = [SCN^-]$$

$$S = \sqrt{1.928 \times 10^{-12}} = 1.4 \times 10^{-6} \text{ M}$$

(2) $S = \sqrt{1.1 \times 10^{-12}} = 1.0 \times 10^{-6} \text{ M}$

(b)

$$PbI_2(s) \rightleftharpoons Pb^{2+} + 2I^-$$

(1) $\gamma_{Pb^{2+}} = 0.36$ $\gamma_{I^-} = 0.75$ $K'_{sp} = a_{Pb^{2+}} a_{I^-}^2 = \gamma_{Pb^{2+}}[Pb^{2+}] \times (\gamma_{I^-}[I^-])^2$

$$[Pb^{2+}][I^-]^2 = \frac{7.9 \times 10^{-9}}{\gamma_{Pb^{2+}}\gamma_{I^-}^2} = \frac{7.9 \times 10^{-9}}{(0.36)(0.75)^2} = 3.90 \times 10^{-8}$$

$$\text{Solubility} = S = [Pb^{2+}] = \frac{1}{2}[I^-]$$

$$S(2S)^2 = 3.90 \times 10^{-8}$$

$$S = \left(\frac{3.90 \times 10^{-8}}{4}\right)^{\frac{1}{3}} = 2.1 \times 10^{-3} \text{ M}$$

(2) $S = \left(\dfrac{7.9 \times 10^{-9}}{4}\right)^{\frac{1}{3}} = 1.3 \times 10^{-3} \text{ M}$

(c) $BaSO_4(s) \rightleftharpoons Ba^{2+} + SO_4^{2-}$

$$\gamma_{Ba^{2+}} = 0.38; \quad \gamma_{SO_4^{2-}} = 0.35$$

$$[Ba^{2+}][SO_4^{2-}] = \frac{1.1 \times 10^{-10}}{\gamma_{Ba^{2+}}\gamma_{SO_4^{2-}}} = \frac{1.1 \times 10^{-10}}{(0.38)(0.35)} = 8.3 \times 10^{-10}$$

$$\text{Solubility} = S = [Ba^{2+}] = [SO_4^{2-}]$$

$$S^2 = 8.3 \times 10^{-10}$$

$$S = \sqrt{8.3 \times 10^{-10}} = 2.9 \times 10^{-5} \text{ M}$$

(2) $S = \sqrt{1.1 \times 10^{-10}} = 1.0 \times 10^{-5} \text{ M}$

(d) $Cd_2Fe(CN)_6(s) \rightleftharpoons 2\,Cd^{2+} + Fe(CN)_6^{4-}$

(1) $\gamma_{Cd^{2+}} = 0.38$ $\gamma_{Fe(CN)_6^{4-}} = 0.020$

$$[Cd^{2+}]^2[Fe(CN)_6^{4-}] = \frac{3.2 \times 10^{-17}}{\gamma_{Cd^{2+}}^2 \gamma_{Fe(CN)_6^{4-}}} = \frac{3.2 \times 10^{-17}}{(0.38)^2(0.020)} = 1.108 \times 10^{-14}$$

$$\text{Solubility} = S = \frac{1}{2}[Cd^{2+}] = [Fe(CN)_6^{4-}]$$

$$(2S)^2 S = 1.108 \times 10^{-14}$$

$$S = \left(\frac{1.108 \times 10^{-14}}{4}\right)^{\frac{1}{3}} = 1.4 \times 10^{-5}\ M$$

(2) $S = (\dfrac{3.2 \times 10^{-17}}{4})^{\frac{1}{3}} = 2.0 \times 10^{-6}\ M$

10-15. $\mu = \frac{1}{2}[0.0167 \times 2^2 + 2 \times 0.0167 \times 1^2] = 0.050$

(a) $AgIO_3(s) \rightleftharpoons Ag^+ + IO_3^-$

(1) $\gamma_{Ag^+} = 0.80$ $\gamma_{IO_3^-} = 0.82$

$$[Ag^+][IO_3^-] = \frac{3.1 \times 10^{-8}}{\gamma_{Ag^+}\gamma_{IO_3^-}} = \frac{3.1 \times 10^{-8}}{(0.80)(0.82)} = 4.7 \times 10^{-8}$$

$$\text{Solubility} = S = [Ag^+] = [IO_3^-]$$

$$S^2 = 4.7 \times 10^{-8}$$

$$S = \sqrt{4.7 \times 10^{-8}} = 2.2 \times 10^{-4}\ M$$

(2) $S = \sqrt{3.1 \times 10^{-8}} = 1.8 \times 10^{-4}\ M$

(b) $Mg(OH)_2(s) \rightleftharpoons Mg^{2+} + 2OH^-$

(1) $\gamma_{Mg^{2+}} = 0.52$ $\gamma_{OH^-} = 0.8$

$$[Mg^{2+}][OH^-]^2 = \frac{7.1 \times 10^{-12}}{\gamma_{Mg^{2+}}\gamma_{OH^-}^2} = \frac{7.1 \times 10^{-12}}{(0.52)(0.81)^2} = 2.081 \times 10^{-11}$$

$$\text{Solubility} = S = [Mg^{2+}] = \frac{1}{2}[OH^-]$$

$$S(2S)^2 = 2.081 \times 10^{-11}$$

$$S = \left(\frac{2.081 \times 10^{-11}}{4}\right)^{\frac{1}{3}} = 1.7 \times 10^{-4}\ M$$

$$(2)\quad S = \left(\frac{7.1 \times 10^{-12}}{4}\right)^{\frac{1}{3}} = 1.2 \times 10^{-4}\ M$$

(c) $BaSO_4(s) \rightleftharpoons Ba^{2+} + SO_4^{2-}$

$$(1)\quad \gamma_{Ba^{2+}} = 0.46 \quad \gamma_{SO_4^{2-}} = 0.44$$

$$[Ba^{2+}][SO_4^{2-}] = \frac{1.1 \times 10^{-10}}{\gamma_{Ba^{2+}}\gamma_{SO_4^{2-}}} = \frac{1.1 \times 10^{-10}}{(0.46)(0.44)} = 5.435 \times 10^{-10}$$

$$\text{Solubility} = S = [SO_4^{2-}]$$

$$(0.0167) \times S = 5.435 \times 10^{-10}$$

$$S = \left(\frac{5.435 \times 10^{-10}}{0.0167}\right) = 3.3 \times 10^{-8}\ M$$

$$(2)\quad S = \left(\frac{1.1 \times 10^{-10}}{0.0167}\right) = 6.6 \times 10^{-9}\ M$$

(d) $La(IO_3)_3(s) \rightleftharpoons La^{3+} + 3IO_3$

$$(1)\quad \gamma_{La^{3+}} = 0.24 \quad \gamma_{IO_3^-} = 0.82 \quad K_{sp} = a_{La^{3+}}a_{IO_3^-}^3 = \gamma_{La^{3+}}[La^{3+}] \times (\gamma_{IO_3^-}[IO_3^-])^3$$

$$[La^{3+}][IO_3^-]^3 = \frac{1.0 \times 10^{-11}}{\gamma_{La^{3+}}\gamma_{IO_3^-}^3} = \frac{1.0 \times 10^{-11}}{(0.24)(0.82)^3} = 7.557 \times 10^{-11}$$

$$\text{Solubility} = S = [La^{3+}] = \frac{1}{3}[IO_3^-]$$

$$S(3S)^3 = 7.557 \times 10^{-11}$$

$$S = \left(\frac{7.557 \times 10^{-11}}{27} \right)^{\frac{1}{4}} = 1.3 \times 10^{-3} \text{ M}$$

(2) $S = \left(\dfrac{1.0 \times 10^{-11}}{27} \right)^{\frac{1}{4}} = 7.8 \times 10^{-4} \text{ M}$

10-16. (a) $CuCl(s) \rightleftharpoons Cu^+ + Cl^-$

If we assume that Cu^+ has an effective diameter of 0.3 like similarly charged cations, then

(1) $\gamma_{Cu^+} = 0.80$ $\gamma_{Cl^-} = 0.80$ $K'_{sp} = a_{Cu^+} a_{Cl^-} = \gamma_{Cu^+}[Cu^+] \times \gamma_{Cl^-}[Cl^-]$

$$[Cu^+][Cl^-] = \frac{1.9 \times 10^{-7}}{\gamma_{Cu^+}\gamma_{Cl^-}} = \frac{1.9 \times 10^{-7}}{(0.80)(0.80)} = 2.969 \times 10^{-7}$$

Solubility $= S = [Cu^+] = [Cl^-]$

$S^2 = 2.969 \times 10^{-7}$

$S = \sqrt{2.969 \times 10^{-7}} = 5.4 \times 10^{-4} \text{ M}$

(2) $S = \sqrt{1.9 \times 10^{-7}} = 4.4 \times 10^{-4} M$

relative error $= \dfrac{\left(4.4 \times 10^{-4} - 5.4 \times 10^{-4} \right)}{5.4 \times 10^{-4}} \times 100\% = -19\%$

(c) $Fe(OH)_3 \rightleftharpoons Fe^{3+} + 3OH^-$

(1) $\gamma_{Fe^{3+}} = 0.24$ $\gamma_{OH^-} = 0.81$ $K'_{sp} = a_{Fe^{3+}} a^3_{OH^-} = \gamma_{Fe^{3+}}[Fe^{3+}] \times (\gamma_{OH^-}[OH^-])^3$

$$[Fe^{3+}][OH^-]^3 = \frac{2 \times 10^{-39}}{\gamma_{Fe^{3+}} \gamma_{OH^-}^3} = \frac{2 \times 10^{-39}}{(0.24)(0.81)^3} = 1.568 \times 10^{-38} \quad \text{retaining figures until the end}$$

$$\text{Solubility} = S = [Fe^{3+}] = \frac{1}{3}[OH^-]$$

$$S(3S)^3 = 1.568 \times 10^{-38}$$

$$S = \left(\frac{1.568 \times 10^{-38}}{27}\right)^{\frac{1}{4}} = 1.55 \times 10^{-10} \text{ M}$$

$$(2) \quad S = \left(\frac{2 \times 10^{-39}}{27}\right)^{\frac{1}{4}} = 9.3 \times 10^{-11} M$$

$$\text{relative error} = \frac{9.3 \times 10^{-11} - 1.55 \times 10^{-10}}{1.55 \times 10^{-10}} \times 100\% = -40\%$$

(e) $Ag_3(AsO_4)(s) \rightleftharpoons 3Ag^+ + AsO_4^{3-}$

Since the α_X of AsO_4^{3-} was given as 0.4, the γ value will be like PO_4^{3-}. So,

$$(1) \quad \gamma_{Ag^+} = 0.80 \quad \gamma_{AsO_4^{3-}} = 0.16 \quad K'_{sp} = a_{Ag^+}^3 a_{AsO_4^{3-}} = (\gamma_{Ag^+}[Ag^+])^3 \times \gamma_{AsO_4^{3-}}[AsO_4^{3-}]$$

$$[Ag^+]^3[AsO_4^{3-}] = \frac{6 \times 10^{-23}}{\gamma_{Ag^+}^3 \gamma_{AsO_4^{3-}}} = \frac{6 \times 10^{-23}}{(0.80)^3(0.16)} = 7.324 \times 10^{-22}$$

$$\text{Solubility} = S = [AsO_3^{4-}] = \frac{1}{3}[Ag^+]$$

$$(3S)^3 S = 7.324 \times 10^{-22}$$

$$S = \left(\frac{7.324 \times 10^{-22}}{27}\right)^{\frac{1}{4}} = 2.3 \times 10^{-6} \text{ M}$$

$$(2) \quad S = \left(\frac{6 \times 10^{-23}}{27}\right)^{\frac{1}{4}} = 1.2 \times 10^{-6} \text{ M}$$

$$\text{relative error} = \frac{1.2 \times 10^{-6} - 2.3 \times 10^{-6}}{2.3 \times 10^{-6}} \times 100\% = -48\%$$

10-17. (a)

In this buffer solution, we assume $[HOAc] = c_{HOAc}$ and $[OAC^-] = c_{NaOAc}$. We also

assume that the ionic strength is contributed solely by NaOAc, neglecting H_3O^+ and OH^-.

$$\mu = \frac{1}{2}[0.250 \times 1^2 + 0.250 \times 1^2] = 0.250$$

$$-\log \gamma_{H_3O^+} = \frac{(0.51)(1)^2 \sqrt{0.250}}{1+(3.3)(0.9)\sqrt{0.250}} = 0.1026 \quad \gamma_{H_3O^+} = 0.790$$

$$-\log \gamma_{OAc^-} = \frac{(0.51)(1)^2 \sqrt{0.250}}{1+(3.3)(0.425)\sqrt{0.520}} = 0.1499 \quad \gamma_{OAc^-} = 0.708$$

$$K_a = \frac{\gamma_{H_3O^+}[H_3O^+]\gamma_{OAc^-}[OAc^-]}{[HOAc]}$$

$$K_a' = \frac{[H_3O^+][OAc^-]}{[HOAc]} = \frac{K_a}{\gamma_{H_3O^+}\gamma_{OAc^-}} = \frac{1.75 \times 10^{-5}}{0.790 \times 0.708} = 3.129 \times 10^{-5}$$

$$[H_3O^+] = \frac{K_a'[HOAc]}{[OAc^-]} = \frac{3.129 \times 10^{-5} \times 0.150}{0.250} = 1.9 \times 10^{-5} \text{ M}$$

pH = 4.73

With no activity corrections

$$[H_3O^+] = \frac{1.75 \times 10^{-5} \times 0.150}{0.250} = 1.05 \times 10^{-5} \text{ M}$$

pH = 4.98

$$\text{relative error in } [H_3O^+] = \frac{1.05 \times 10^{-5} - 1.9 \times 10^{-5}}{1.9 \times 10^{-5}} \times 100\% = -45\%$$

Chapter 11

11-2. To simplify equilibrium calculations, we sometimes assume that the concentrations of

one or more species are negligible and can be approximated as 0.00 M. In a sum or difference

assuming a concentration is 0.00 M leads to an appropriate result. In contrast, if we were to

simplify and equilibrium constant expression by assuming on or more concentrations are zero,

we would be multiplying or dividing by 0.00, which would render the expression meaningless.

11-4. A charge-balance equation is derived by relating the concentration of cations and anions

no. mol/L positive charge = no. mol/L negative charge

For a doubly charged ion, such as Ba^{2+}, the concentration of charge for each mole is

twice the molar *concentration* of the Ba^{2+}. That is,

mol/L positive charge = $2[Ba^{2+}]$

Thus, the molar concentration of all multiply charged species is always multiplied by the

charge in a charge-balance equation.

11-5. **(a)** $0.20 = [HF] + [F^-]$

    (c) $0.10 = [H_3PO_4] + [H_2PO_4^-] + [HPO_4^{2-}] + [PO_4^{3-}]$

    (e) $0.0500 + 0.100 = [HClO_2] + [ClO_2^-]$

        $[Na^+] = c_{NaClO2} = 0.100$ M

    (g) $0.100 = [Na^+] = [OH^-] + 2 [Zn(OH)_4^{2-}]$

    (i) $[Pb^{2+}] = \frac{1}{2}([F^-] + [HF])$

11-7. Following the systematic procedure, using part (a)

Step 1 $SrC_2O_4(s) \rightleftharpoons Sr^{2+} + C_2O_4^{2-}$

$$H_2C_2O_4 + H_2O \rightleftharpoons H_3O^+ + HC_2O_4^-$$

$$HC_2O_4^- + H_2O \rightleftharpoons H_3O^+ + C_2O_4^{2-}$$

Step 2 $S = $ solubility $= [Sr^{2+}] = [C_2O_4^{2-}] + [HC_2O_4^-] + [H_2C_2O_4]$

Step 3 $[Sr^{2+}][C_2O_4^{2-}] = K_{sp} = 5 \times 10^{-8}$ (1)

$$\frac{[H_3O^+][HC_2O_4^-]}{[H_2C_2O_4]} = K_1 = 5.6 \times 10^{-2} \qquad (2)$$

$$\frac{[H_3O^+][C_2O_4^-]}{[HC_2O_4^-]} = K_2 = 2.42 \times 10^{-5} \qquad (3)$$

Step 4 $[Sr^{2+}] = [C_2O_4^{2-}] + [HC_2O_4^-] + [H_2C_2O_4]$ (4)

$$[H_3O^+] = 1.0 \times 10^{-6} \text{ M}$$

Step 5 No charge balance because an unknown buffer is maintaining the pH.

Step 6 Unknowns are $[Sr^{2+}]$, $[C_2O_4^{2-}]$, $[HC_2O_4^-]$, $[H_2C_2O_4]$

Step 7 No approximations needed, because we have 4 equations and 4 unkowns.

Step 8 Substituting $[H_3O^+] = 1.0 \times 10^{-6}$ M into equation (3) and rearranging gives

$$[HC_2O_4^-] = \frac{1 \times 10^{-6}[C_2O_4^-]}{5.42 \times 10^{-5}} = 1.845 \times 10^{-2}[C_2O_4^-]$$

Substituting this relationship and $[H_3O^+] = 1.0 \times 10^{-6}$ M into equation (2) and rearranging

gives

$$[H_2C_2O_4] = \frac{1 \times 10^{-6} \times 1.845 \times 10^{-2}[C_2O_4^-]}{5.6 \times 10^{-2}} = 3.295 \times 10^{-7}[C_2O_4^-]$$

64

Substituting these last two relationships in to equation (4) gives

$$[Sr^{2+}] = [C_2O_4^{2-}] + 1.845 \times 10^{-2}[C_2O_4^{2-}] + 3.295 \times 10^{-7}[C_2O_4^{2-}] = 1.0185[C_2O_4^{2-}]$$

Substituting this last relationship into equation (1) gives

$$K_{sp} = \frac{[Sr^{2+}][Sr^{2+}]}{1.0185} = 5 \times 10^{-8}$$

$$[Sr^{2+}] = (5 \times 10^{-8} \times 1.0185)^{1/2} = 2.26 \times 10^{-4}$$

$$S = [Sr^{2+}] = 2.3 \times 10^{-4} \text{ M}$$

Substituting other values for $[H_3O^+]$ gives the following:

	$[H_3O^+]$	S, M
(a)	1.00×10^{-6}	2.3×10^{-4}
(c)	1.00×10^{-9}	2.2×10^{-4}

11-8.   Proceeding as in Problem 11-7, we write

$$BaSO_4 \rightleftharpoons Ba^{2+} + SO_4^{2-} \qquad\qquad K_{sp} = 1.1 \times 10^{-10}$$

$$HSO_4^- + H_2O \rightleftharpoons H_3O^+ + SO_4^{2-} \qquad K_2 = 1.02 \times 10^{-2}$$

$$S = [Ba^{2+}]$$

$$[Ba^{2+}][SO_4^{2-}] = 1.1 \times 10^{-10} \qquad\qquad\qquad (1)$$

$$\frac{[H_3O^+][SO_4^{2-}]}{[HSO_4^-]} = 1.02 \times 10^{-2} \qquad\qquad\qquad (2)$$

Mass balance requires that

$$[Ba^{2+}] = [SO_4^{2-}] + [HSO_4^-] \qquad\qquad\qquad (3)$$

The unknowns are $[Ba^{2+}]$, $[SO_4^{2-}]$, and $[HSO_4^-]$

We have 3 equations and 3 unknowns so no approximations are needed.

Substituting eqation (2) into (3) gives

$$[Ba^{2+}] = [SO_4^{2-}] + \frac{[H_3O^+][SO_4^{2-}]}{1.02 \times 10^{-2}} = [SO_4^{2-}]\left(1 + \frac{[H_3O^+]}{1.02 \times 10^{-2}}\right)$$

Substituting equation (1) to eliminate $[SO_4^{2-}]$, gives

$$[Ba^{2+}] = \frac{1.1 \times 10^{-10}}{[Ba^{2+}]} \times \left(1 + \frac{[H_3O^+]}{1.02 \times 10^{-2}}\right) = \frac{1.1 \times 10^{-10}}{[Ba^{2+}]} \times \left(1 + 98.0[H_3O^+]\right)$$

$$S = [Ba^{2+}] = \sqrt{1.1 \times 10^{-10}\left(1 + 98.0[H_3O^+]\right)} = \sqrt{1.1 \times 10^{-10} + 1.078 \times 10^{-8}[H_3O^+]}$$

Using the different values of $[H_3O^+]$

	$[H_3O^+]$	S, M
(a)	3.5	1.9×10^{-4}
(c)	0.08	3.1×10^{-5}

11-9. The derivation that follows applies to problems 9-11.

$$MS(s) \rightleftharpoons M^{2+} + S^{2-} \qquad\qquad K_{sp}$$

$$H_2S + H_2O \rightleftharpoons H_3O^+ + HS^- \qquad K_1 = 9.6 \times 10^{-8}$$

$$HS^- + H_2O \rightleftharpoons H_3O^+ + S^{2-} \qquad K_2 = 1.3 \times 10^{-14}$$

Overall $H_2S + 2H_2O \rightleftharpoons 2H_3O^+ + S^{2-} \qquad K_1K_2 = 1.25 \times 10^{-21}$

$$S = \text{solubility} = [M^{2+}]$$

$$[M^{2+}][S^{2-}] = K_{sp} \qquad\qquad (1)$$

$$\frac{[H_3O^+][HS^-]}{[H_2S]} = K_2 = 1.3 \times 10^{-14} \qquad (2)$$

$$\frac{[H_3O^+]^2[S^{2-}]}{[H_2S]} = K_1K_2 = 1.25 \times 10^{-21} \qquad (3)$$

Mass balance is:

$$[M^{2+}] = [S^{2-}] + [HS^-] + [H_2S] \qquad (4)$$

Substituting equation (2) and (3) into (4), gives:

$$[M^{2+}] = [S^{2-}] + \frac{[H_3O^+][S^{2-}]}{K_2} + \frac{[H_3O^+]^2[S^{2-}]}{K_1K_2} = [S^{2-}]\left(1 + \frac{[H_3O^+]}{K_2} + \frac{[H_3O^+]^2}{K_1K_2}\right) \qquad (5)$$

Substituting equation (1) into (5), gives

$$[M^{2+}] = \frac{K_{sp}}{[M^{2+}]}\left(1 + \frac{[H_3O^+]}{K_2} + \frac{[H_3O^+]^2}{K_1K_2}\right)$$

$$[M^{2+}] = \sqrt{K_{sp}\left(1 + \frac{[H_3O^+]}{1.3 \times 10^{-14}} + \frac{[H_3O^+]^2}{1.25 \times 10^{-21}}\right)} \qquad (6)$$

(a) Substituting $K_{sp} = 3 \times 10^{-28}$ and $[H_3O^+] = 3.0 \times 10^{-1}$ into equation (6), gives

$$[M^{2+}] = \text{solubility} = \sqrt{3 \times 10^{-28} \left(1 + \frac{0.30}{1.3 \times 10^{-14}} + \frac{(0.30)^2}{1.25 \times 10^{-21}} \right)} = 1.5 \times 10^{-4} \text{ M}$$

(b) Using the same K_{sp}, but $[H_3O^+] = 3.0 \times 10^{-4}$, gives

$$[M^{2+}] = \text{solubility} = 1.5 \times 10^{-7} \text{ M}$$

11-11. For MnS(pink), $K_{sp} = 3.0 \times 10^{-11}$

(a) For $[H_3O^+] = 3.00 \times 10^{-5}$

$$[M^{2+}] = \text{solubility} = 4.7 \text{ M}$$

11-12. Proceeding as in Problem 11-9, we find

$$[Zn^{2+}] = \sqrt{K_{sp} \left(1 + \frac{[H_3O^+]}{K_2} + \frac{[H_3O^+]^2}{K_1 K_2} \right)}$$

For ZnCO$_3$, $K_{sp} = 1.0 \times 10^{-10}$. For H$_2CO_3$, $K_1 = 4.45 \times 10^{-7}$, and $K_2 = 4.69 \times 10^{-11}$

$$[Zn^{2+}] = \sqrt{1 \times 10^{-10} \left(1 + \frac{[H_3O^+]}{4.69 \times 10^{-11}} + \frac{[H_3O^+]^2}{4.45 \times 10^{-7} \times 4.69 \times 10^{-11}} \right)}$$

For pH = 7.00, $[H_3O^+] = 1.00 \times 10^{-7}$

$$[Zn^{2+}] = 5.1 \times 10^{-4} \text{ M}$$

11-14. $[Cu^{2+}][OH^-]^2 = 4.8 \times 10^{-20}$ $[Mn^{2+}][OH^-]^2 = 2 \times 10^{-13}$

(a)   Cu(OH)$_2$ precipitates first

(b)   Cu^{2+} begins to precipitate when

$$[OH^-] = \sqrt{\frac{4.8 \times 10^{-20}}{0.05}} = 9.8 \times 10^{-10} \text{ M}$$

(c) Mn^{2+} begins to precipitate when

$$[OH^-] = \sqrt{\frac{2 \times 10^{-13}}{0.04}} = 2.24 \times 10^{-6} \text{ M}$$

$$[Cu^{2+}] = 4.8 \times 10^{-20}/(2.24 \times 10^{-6})^2 = 9.6 \times 10^{-9} \text{ M}$$

11-16. (a) $[Ag^+] = K_{sp}/[I^-] = 8.3 \times 10^{-17}/(1.0 \times 10^{-6}) = 8.3 \times 10^{-11} \text{ M}$

(b) $[Ag^+] = K_{sp}/[SCN-] = 1.1 \times 10^{-12}/(0.080) = 1.375 \times 10^{-11} \text{ M} \approx 1.4 \times 10^{-11} \text{ M}$

(c) $[I^-]$ when $[Ag^+] = 1.375 \times 10^{-11} \text{ M}$

$$[I^-] = 8.3 \times 10^{-17}/(1.375 \times 10^{-11}) = 6.0 \times 10^{-6} \text{ M}$$

$$[SCN^-]/[I^-] = 0.080/(6.0 \times 10^{-6}) = 1.3 \times 10^4$$

(d) $[I^-] = 8.3 \times 10^{-17}/(1.0 \times 10^{-3}) = 8.3 \times 10^{-14} \text{ M}$

$$[SCN^-] = 1.1 \times 10^{-12}/(1.0 \times 10^{-3}) = 1.1 \times 10^{-9} \text{ M}$$

$$[SCN^-]/[I^-] = 1.1 \times 10^{-9}/(8.3 \times 10^{-14}) = 1.3 \times 10^4$$

Note that this ratio is independent of $[Ag^+]$ as long as some $AgSCN_{(s)}$ is present.

11-18. $AgBr \rightleftharpoons Ag^+ + Br^-$ $K_{sp} = 5.0 \times 10^{-13} = [Ag^+][Br^-]$ (1)

$$Ag^+ + 2CN^- \rightleftharpoons Ag(CN)_2^- \qquad \beta_2 = 1.3 \times 10^{21} = \frac{[Ag(CN)_2^-]}{[Ag^+][CN^-]^2} \qquad (2)$$

It is readily shown that $CN^- + H_2O \rightleftharpoons HCN + OH^-$ can be neglected in this problem so

that only the two equilibria shown above need to be considered.

Solubility $= [Br^-]$

Mass balance requires that

$[Br^-] = [Ag^+] + [Ag(CN)_2^-]$ (3)

$$0.200 = [CN^-] + 2[Ag(CN)_2^-] \qquad (4)$$

We now have 4 equations and 4 unknowns.

Because β_2 is very large, let us assume that

$$[CN^-] \ll 2[Ag(CN)_2^-] \qquad \text{and} \qquad [Ag^+] \ll [Ag(CN)_2^-]$$

(4) becomes $[Ag(CN)_2^-] = 0.200/2 = 0.100$

and (3) becomes $[Br^-] = [Ag(CN)_2^-] = 0.100$

To check the assumptions, we calculate $[Ag^+]$ by substituting into (1)

$$[Ag^+] = 5.0 \times 10^{-13}/0.100 \cong 5 \times 10^{-12} \qquad (5 \times 10^{-12} \ll$$

$0.100)$

To obtain $[CN^-]$ we substitute into (2) and rearrange

$$[CN^-] = \sqrt{\frac{0.100}{\left(1 \times 10^{-11}\right)\left(1.3 \times 10^{21}\right)}} = 2.8 \times 10^{-6} \qquad (2.8 \times 10^{-6} \ll 0.100)$$

Thus, the two assumptions are valid and

$$\text{Solubility} = [Br^-] = 0.100 \text{ M}$$

$$\text{mass AgBr}/200 \text{ mL} = 0.100 \frac{\text{mmol}}{\text{mL}} \times 200 \text{ mL} \times \frac{0.1877 \text{ g}}{\text{mmol}}$$

$$= 3.754 \text{ g}$$

11-20. $\qquad CaSO_{4(s)} \rightleftharpoons Ca^{2+} + SO_4^{2-} \qquad K_{sp} = [Ca^{2+}][SO_4^{2-}] = 2.6 \times 10^{-5} \qquad (1)$

$$CaSO_{4(aq)} \rightleftharpoons Ca^{2+} + SO_4^{2-} \qquad K_d = \frac{[Ca^{2+}][SO_4^-]}{[CaSO_4]_{aq}} = 5.2 \times 10^{-3} \qquad (2)$$

$$CaSO_{4(s)} \rightleftharpoons CaSO_{4(aq)} \qquad (3)$$

The mass balance gives

$$[Ca^{2+}] = [SO_4^{2-}] \qquad (4)$$

We have 3 equations and 3 unknowns ($[Ca^{2+}]$, $[SO_4^{2-}]$, and $[CaSO_4]_{aq}$

To solve we divide (1) by (2) to give

$$[CaSO_4]_{aq} = K_{sp}/K_d = (2.6 \times 10^{-5})/(5.2 \times 10^{-3}) = 5.0 \times 10^{-3}$$

Note that this is the equilibrium constant expression for (3) and indicates that the

concentration of un-ionized $CaSO_4$ is always the same in a saturated solution of $CaSO_4$.

Substituting (4) into (1) gives

$$[Ca^{2+}] = (2.6 \times 10^{-5})^{1/2} = 5.1 \times 10^{-3} \text{ M}$$

and since $S = [CaSO_4]_{aq} + [Ca^{2+}]$, we obtain

$$S = 5.0 \times 10^{-3} + 5.1 \times 10^{-3} = 1.01 \times 10^{-2} \text{ M}$$

$$\%CaSO_{4(aq)} = (5.0 \times 10^{-3}/1.01 \times 10^{-2}) \times 100\% = 49\%$$

(b) Here $[CaSO_4]_{aq}$ is again equal to 5.0×10^{-3} and the mass balance gives

$$[SO_4^{2-}] = 0.0100 + [Ca^{2+}] \qquad (5)$$

Substituting (1) into (5) and rearranging gives

$$0 = [SO_4^{2-}]^2 - 0.0100[SO_4^{2-}] - K_{sp}$$

which may be solved using the quadratic equation to give

$$[SO_4^{2-}] = 0.0121 \quad \text{and} \quad [Ca^{2+}] = 2.14 \times 10^{-3}$$

$$S = 5.0 \times 10^{-3} + 2.14 \times 10^{-3} = 7.14 \times 10^{-3} \text{ M}$$

$$\%CaSO_{4(aq)} = (5.0 \times 10^{-3}/7.14 \times 10^{-3}) \times 100\% = 70\%$$

Chapter 12

12-1. **(a)** A *colloidal precipitate* consists of solid particles with dimensions that are less than

10^{-4} cm. A *crystalline precipitate* consists of solid particles with dimensions that at least

10^{-4} cm or greater. As a result, crystalline precipitates settle rapidly, whereas colloidal

precipitates remain suspended in solution unless caused to agglomerate.

(c) *Precipitation* is the process by which a solid phase forms and is carried out of

solution when the solubility product of a chemical species is exceeded. *Coprecipitation*

is a process in which normally soluble compounds are carried out of solution during

precipitate formation.

(e) *Occlusion* is a type of coprecipitation in which a compound is trapped within a

pocket formed during rapid crystal formation. *Mixed-crystal formation* is also a type of

coprecipitation in which a contaminant ion replaces an ion in the crystal lattice.

12-2. **(a)** *Digestion* is a process in which a precipitate is heated in the presence of the solution

from which it was formed (the *mother liquor*). Digestion improves the purity and

filterability of the precipitate.

(c) In *reprecipitation*, the filtered solid precipitate is redissolved and reprecipitated.

Because the concentration of the impurity in the new solution is lower, the second

precipitate contains less coprecipitated impurity.

(e) The *counter-ion layer* describes a layer of solution containing sufficient excess

negative ions that surrounds a charged particle. This counter-ion layer balances the

surface charge on the particle.

(g) *Supersaturation* describes an unstable state in which a solution contains higher solute concentration than a saturated solution. Supersaturation is relieved by precipitation of excess solute.

12-3.　A *chelating agent* is an organic compound that contains two or more electron-donor groups located in such a configuration that five- or six-membered rings are formed when the donor groups complex a cation.

12-5.　(a) There is positive charge on the surface of the coagulated colloidal particles.

(b) The positive charge arises from adsorbed Ag^+ ions.

(c) NO_3^- ions make up the counter-ion layer.

12-7.　In *peptization*, a coagulated colloid returns to its original dispersed state because of a decrease in the electrolyte concentration of the solution contacting the precipitate. Peptization can be avoided by washing the coagulated colloid with an electrolyte solution instead of pure water.

12-9.　(a) mass SO_2 = mass $BaSO_4 \times \dfrac{\mathcal{M}_{SO_2}}{\mathcal{M}_{BaSO_4}}$

(c) mass In = mass $In_2O_3 \times \dfrac{2\mathcal{M}_{In}}{\mathcal{M}_{In_2O_3}}$

(e) mass CuO = mass $Cu_2(SCN)_2 \times \dfrac{2\mathcal{M}_{CuO}}{\mathcal{M}_{Cu_2(SCN)_2}}$

(i) mass $Na_2B_4O_7 \cdot 10H_2O$ = mass $B_2O_3 \times \dfrac{\mathcal{M}_{Na_2B_4O_7 \cdot 10H_2O}}{2\mathcal{M}_{B_2O_3}}$

12-10. $\mathcal{M}_{AgCl} = 143.32$ g/mol $\mathcal{M}_{KCl} = 74.55$ g/mol

$$\frac{0.2912 \text{ g AgCl} \times \left(\dfrac{1 \text{ mol AgCl}}{143.32 \text{ g}}\right) \times \left(\dfrac{1 \text{ mol KCl}}{1 \text{ mol AgCl}}\right) \times \left(\dfrac{74.55 \text{ g KCl}}{\text{mol}}\right)}{0.2500 \text{ g impure sample}} \times 100\% = 60.59\%$$

12-12.

$$0.650 \text{ g CuSO}_4 \cdot 5\text{H}_2\text{O} \times \frac{1 \text{ mol CuSO}_4 \cdot 5\text{H}_2\text{O}}{249.67 \text{ g CuSO}_4 \cdot 5\text{H}_2\text{O}} \times \frac{1 \text{ mol Cu(IO}_3)_2}{1 \text{ mol CuSO}_4 \cdot 5\text{H}_2\text{O}}$$

$$\times \frac{413.35 \text{ g Cu(IO}_3)_2}{1 \text{ mol Cu(IO}_3)_2} = 1.076 \text{ g Cu(IO}_3)_2$$

12-14.

$$\text{mass AgI} = 0.512 \text{ g} \times \frac{20.1 \text{ g}}{100 \text{ g}} \times \frac{1 \text{ mol AlI}_3}{407.69 \text{ g}} \times \frac{3 \text{ mol AgI}}{\text{mol AlI}_3} \times \frac{234.77 \text{ g AgI}}{\text{mol AgI}} = 0.178 \text{ g}$$

12-18. $$\frac{0.6006 \text{ g BaCO}_3 \times \dfrac{1 \text{ mol BaCO}_3}{197.34 \text{ g}} \times \dfrac{1 \text{ mol C}}{1 \text{ mol BaCO}_3} \times \dfrac{12.011 \text{ g C}}{1 \text{ mol C}}}{0.2121 \text{ g sample}} \times 100\% = 17.23\% \text{ C}$$

12-20.

$$\frac{\left(\begin{array}{l} 0.5718 \text{ g Hg}_5(\text{IO}_6)_2 \times \dfrac{1 \text{ mol Hg}_5(\text{IO}_6)_2}{1448.75 \text{ g Hg}_5(\text{IO}_6)_2} \times \dfrac{5 \text{ mol Hg}^{2+}}{1 \text{ mol Hg}_5(\text{IO}_6)_2} \\ \times \dfrac{1 \text{ mol Hg}_2\text{Cl}_2}{2 \text{ mol Hg}^{2+}} \times \dfrac{472.18 \text{ g Hg}_2\text{Cl}_2}{1 \text{ mol Hg}_2\text{Cl}_2} \end{array}\right)}{1.0451 \text{ g sample}} \times 100\% = 44.58\% \text{ Hg}_2\text{Cl}_2$$

12-22. $\mathcal{M}_{NH3} = 17.0306$ g/mol $\mathcal{M}_{Pt} = 195.08$ g/mol

$$\frac{0.4693 \text{ g Pt} \times \left(\dfrac{1 \text{ mol Pt}}{195.08 \text{ g}}\right) \times \left(\dfrac{2 \text{ mol NH}_3}{1 \text{ mol Pt}}\right) \times \left(\dfrac{17.0306 \text{ g NH}_3}{\text{mol}}\right)}{0.2115 \text{ g impure sample}} \times 100\% = 38.74\% \text{ NH}_3$$

12-24. $\mathcal{M}_{BaSO4} = 233.39$ g/mol　　　　$\mathcal{M}_{SO42-} = 96.064$ g/mol

Let S_w = mass of sample in grams.

$$0.200 \text{ g BaSO}_4 \times \frac{1 \text{ mol BaSO}_4}{233.39 \text{ g}} \times \frac{1 \text{ mol SO}_4^{2-}}{1 \text{ mol BaSO}_4} = 8.57 \times 10^{-4} \text{ mol SO}_4^{2-}$$

$$\frac{8.57 \times 10^{-4} \text{ mol SO}_4^{2-} \times \dfrac{96.064 \text{ g SO}_4^{2-}}{\text{mol}}}{S_w \text{ g sample}} = 100\% = 20\% \text{ SO}_4^{2-}$$

$$S_w = \frac{8.57 \times 10^{-4} \text{ mol SO}_4^{2-} \times \dfrac{96.064 \text{ g SO}_4^{2-}}{\text{mol}} \times 100\%}{20\%} = 0.412 \text{ g sample}$$

The maximum precipitate mass expected given this sample mass is

$$0.412 \text{ g sample} \times \frac{55 \text{ g SO}_4^{2-}}{100 \text{ g sample}} \times \frac{1 \text{ mol SO}_4^{2-}}{96.064 \text{ g}} \times \frac{1 \text{ mol BaSO}_4}{1 \text{ g SO}_4^{2-}} \times \frac{233.39 \text{ g BaSO}_4}{1 \text{ mol}}$$

$$= 0.550 \text{ g BaSO}_4$$

12-26. Let S_w = mass of sample in grams.

(a)　　$\mathcal{M}_{AgCl} = 143.32$ g/mol　　　　$\mathcal{M}_{ZrCl4} = 233.03$ g/mol

$$\frac{0.400 \text{ g AgCl} \times \dfrac{1 \text{ mol AgCl}}{143.32 \text{ g}} \times \dfrac{1 \text{ mol ZrCl}_4}{4 \text{ mol AgCl}} \times \dfrac{233.03 \text{ g ZrCl}_4}{1 \text{ mol}}}{S_w \text{ g sample}} \times 100\% = 68\% \text{ ZrCl}_4$$

$$S_w = \frac{1.62 \times 10^{-1} \text{ g ZrCl}_4 \times 100\%}{68\%} = 0.239 \text{ g sample}$$

(b)

$$0.239 \text{ g sample} \times \frac{84 \text{ g ZrCl}_4}{100 \text{ g sample}} \times \frac{1 \text{ mole ZrCl}_4}{233.03 \text{ g}} \times \frac{4 \text{ mole AgCl}}{1 \text{ mole ZrCl}_4} \times \frac{143.32 \text{ g AgCl}}{1 \text{ mole}} = 0.494 \text{ g AgCl}$$

(c)

$$\% \text{ ZrCl}_4 = \frac{1.62 \times 10^{-1} \text{ g ZrCl}_4 \times 100\%}{S_w} = 40\%$$

$$S_w = \frac{1.62 \times 10^{-1} \text{ g ZrCl}_4 \times 100\%}{40\%} = 0.406 \text{ g sample}$$

12-28. $\mathcal{M}_{AgCl} = 143.32$ g/mol $\mathcal{M}_{AgI} = 234.77$ g/mol

0.4430 g = x g AgCl + y g AgI

$$g\ AgCl = x\ g\ AgCl + \left(y\ g\ AgI \times \frac{1\ mol\ AgI}{234.77\ g} \times \frac{1\ mol\ AgCl}{1\ mol\ AgI} \times \frac{143.32\ g\ AgCl}{1\ mol} \right) = 0.3181\ g$$

0.3181 = x g AgCl + 0.6104698 y g AgI

Here again, we have 2 equations and 2 unknowns,

$x + y = 0.4430$

$x + 0.6104698y = 0.3181$

The spreadsheet is shown on the next page

We would report %Cl = 4.72 and %I = 27.05

▲	A	B	C	D
1	**Problem 12-28**			
2	$\mathcal{M}_{AgCl}$	143.32		
3	$\mathcal{M}_{AgI}$	243.77		
4	$\mathcal{M}_{Cl}$	35.453		
5	$\mathcal{M}_{I}$	126.9045		
6				
7	**Coefficient Matrix**			**Constant Matrix**
8	1	1		0.443
9	1	0.6104698		0.3181
10				
11	**Inverse Matrix**			**Solution Matrix**
12	-1.567195047	2.56719505		0.122357339
13	2.567195047	-2.56719505		0.320642661
14				
15	**Sample Mass**	0.6407		
16	**Mass AgCl**	0.12235734		
17	**Mass AgI**	0.32064266		
18	**%Cl**	4.72412619		
19	**%I**	26.0533363		
20				
21	**Documentation**			
22	Cells A12:B13=MINVERSE(A8:B9)			
23	Cells D12:D13=MMULT(A12:B13,D8:D9)			
24	Cell B18=(B16*B4/B2)/B15*100			
25	Cell B19=(B17*B5/B3)/B15*100			

12-30.

$$\mathcal{M}_{CO_2} = 44.010 \text{ g/mol} \quad \mathcal{M}_{MgCO_3} = 84.31 \text{ g/mol} \quad \mathcal{M}_{K_2CO_3} = 138.21 \text{ g/mol}$$

$$\text{mol CO}_2 = \text{mol MgCO}_3 + \text{mol K}_2\text{CO}_3$$

$$= \left(2.300 \text{ g sample} \times \frac{38 \text{ g MgCO}_3}{100 \text{ g sample}} \times \frac{1 \text{ mol MgCO}_3}{84.31 \text{ g}} \right) +$$

$$\left(2.300 \text{ g sample} \times \frac{42 \text{ g K}_2\text{CO}_3}{100 \text{ g sample}} \times \frac{1 \text{ mol K}_2\text{CO}_3}{138.21 \text{ g}} \right)$$

$$\text{amount CO}_2 = 0.0104 + 6.989 \times 10^{-3} = 0.01736 \text{ mol}$$

$$\text{mass CO}_2 = 0.01736 \text{ mole} \times \frac{44.010 \text{ g CO}_2}{1 \text{ mole}} = 0.764 \text{ g}$$

12-32.

$$\mathcal{M}_{BaCl_2 \cdot 2H_2O} = 244.26 \text{ g/mol} \quad \mathcal{M}_{NaIO_3} = 197.89 \text{ g/mol} \quad \mathcal{M}_{Ba(IO_3)_2} = 487.13 \text{ g/mol}$$

$$0.200 \text{ g BaCl}_2 \cdot 2H_2O \times \frac{1 \text{ mol BaCl}_2 \cdot 2H_2O}{244.26 \text{ g}} \times \frac{1 \text{ mol Ba}^{2+}}{1 \text{ mol BaCl}_2 \cdot 2H_2O}$$

$$= 8.188 \times 10^{-4} \text{ mol Ba}^{2+}$$

$$0.300 \text{ g NaIO}_3 \times \frac{1 \text{ mol NaIO}_3}{197.89 \text{ g}} \times \frac{1 \text{ mol IO}_3^-}{1 \text{ mol NaIO}_3} = 1.516 \times 10^{-3} \text{ mol IO}_3^-$$

Because IO_3^- is the limiting reagent,

(a)

$$\text{amount Ba(IO}_3)_2 = \frac{1.516 \times 10^{-3} \text{ mol}}{2} = 7.580 \times 10^{-4} \text{ mol}$$

$$\text{mass Ba(IO}_3)_2 = 7.580 \times 10^{-4} \text{ mol} \times \frac{487.13 \text{ g Ba(IO}_3)_2}{1 \text{ mol}} = 0.369 \text{ g Ba(IO}_3)_2$$

(b)

$$\text{amount BaCl}_2 \cdot 2H_2O \text{ remaining} = \left((8.188 \times 10^{-4}) - (7.580 \times 10^{-4})\right) \text{ mol} = 6.080 \times 10^{-5} \text{ mol}$$

$$\text{mass BaCl}_2 \cdot 2H_2O = 6.08 \times 10^{-5} \text{ mol BaCl}_2 \cdot 2H_2O \times \frac{244.26 \text{ g BaCl}_2 \cdot 2H_2O}{1 \text{ mol}}$$

$$= 0.0149 \text{ g}$$

Chapter 13

13-1. (a) The *millimole* is the amount of an elementary species, such as an atom, an ion, a

molecule, or an electron. A millimole contains

$$6.02 \times 10^{23} \frac{particles}{mol} \times \frac{mol}{1000\ mmol} = 6.02 \times 10^{20} \frac{particles}{mmol}$$

(c) The *stoichiometric ratio* is the molar ratio of two chemical species that appear in a

balanced chemical equation.

13-3. (a) The *equivalence point* in a titration is that point at which sufficient titrant has been

added so that stoichiometrically equivalent amounts of analyte and titrant are present.

The *end point* in a titration is the point at which an observable physical change signals the

equivalence point.

13-5. (a) $\dfrac{1\ mol\ H_2NNH_2}{2\ mol\ I_2}$

(c) $\dfrac{1\ mole\ Na_2B_4O_7 \cdot 10H_2O}{2\ moles\ H^+}$

13-7. (a) $2.95\ mL \times \dfrac{0.0789\ mmol}{mL} = 0.233\ mmol$

(b) $0.2011\ L \times \dfrac{0.0564\ mol}{L} \times \dfrac{1000\ mmol}{mol} = 11.34\ mmol$

(c)

$$\frac{47.5\ g\ Mg(NO_3)_2}{10^6\ g\ soln} \times \frac{1000\ g\ soln}{L} \times \frac{1\ mol}{148.31\ g\ Mg(NO_3)_2} \times 2.56\ L \times \frac{1000\ mmol}{mol} = 0.820\ mmol$$

(d) $79.8\ mL \times \dfrac{0.1379\ mmol}{mL} = 11.00\ mmol$

13-9. **(a)** $450.0 \text{ mL} \times \dfrac{0.0986 \text{ mol } H_2O_2}{L} \times \dfrac{34.02 \text{ g}}{\text{mol } H_2O_2} \times \dfrac{1 \text{ L}}{1000 \text{ mL}} = 1.51 \text{ g}$

(b) $26.4 \text{ mL} \times \dfrac{9.36 \times 10^{-4} \text{ mol}}{L} \times \dfrac{122.1 \text{ g}}{\text{mol}} \times \dfrac{1 \text{ L}}{1000 \text{ mL}} = 3.02 \times 10^{-3} \text{ g}$

(c) $2.50 \text{ L} \times \dfrac{23.4 \text{ mg}}{L} \times \dfrac{1 \text{ g}}{1000 \text{ mg}} = 0.0585 \text{ g}$ (1 ppm = 1 mg/L)

(d) $21.7 \text{ mL} \times \dfrac{0.0214 \text{ mol}}{L} \times \dfrac{167.0 \text{ g}}{\text{mol}} \times \dfrac{1 \text{ L}}{1000 \text{ mL}} = 0.0776 \text{ g}$

13-11. $\dfrac{20.0 \text{ g KCl}}{100 \text{ g soln}} \times \dfrac{1.13 \text{ g soln}}{\text{mL}} \times \dfrac{1 \text{ mmol KCl}}{0.07455 \text{ g KCl}} = 3.03 \dfrac{\text{mmol KCl}}{\text{mL}} = 3.03 \text{ M}$

13-13. **(a)** $1.00 \text{ L} \times \dfrac{0.150 \text{ mol}}{L} \times \dfrac{158.03 \text{ g}}{\text{mol}} = 23.70 \text{ g}$

Dissolve 23.70 g KMnO$_4$ in water and dilute to 1.00 L total volume.

(b) 2.50 L of 0.500 M HClO$_4$ contains $2.50 \text{ L} \times \dfrac{0.500 \text{ mol}}{L} = 1.25 \text{ mol}$

Need to take a volume of $\dfrac{1.25 \text{ mol}}{9.00 \text{ mol}/L} = 0.139 \text{ L}$

Take 139 mL of concentrated (9.00 M) reagent and dilute to 2.50 L.

(c)

$400 \text{ mL} \times \dfrac{0.0500 \text{ mol } I^-}{L} \times \dfrac{1 \text{ L}}{1000 \text{ mL}} \times \dfrac{1 \text{ mol } MgI_2}{2 \text{ mol } I^-} \times \dfrac{278.11 \text{ g}}{\text{mol } MgI_2} = 2.78 \text{ g}$

Dissolve 2.78 g MgI$_2$ in water and bring to 400 mL total volume.

(d)

$$200 \ \cancel{mL} \times \frac{1.00 \ \text{g} \ \cancel{CuSO_4}}{100 \ \cancel{mL}} \times \frac{1 \ \cancel{mol}}{159.61 \ \text{g} \ \cancel{CuSO_4}} \times \frac{1 \ \cancel{L}}{0.218 \ \cancel{mol}} \times \frac{1000 \ \text{mL}}{1 \ \cancel{L}} = 57.5 \ \text{mL}$$

Take 57.5 mL of the 0.218 M solution and dilute to a volume of 200 mL.

(e)   In 1.50 L of 0.215 M NaOH, there are

$$\frac{0.215 \ \text{mole NaOH}}{L} \times 1.50 \ \text{L} = 0.3225 \ \text{mole NaOH}$$

The commercial reagent is $\dfrac{1.525 \times 10^3 \ \cancel{g}}{L} \times \dfrac{50 \ \text{g} \ \cancel{NaOH}}{100 \ \cancel{g}} \times \dfrac{\text{mole}}{40.00 \ \text{g} \ \cancel{NaOH}} = 19.06 \ \text{M}$

Thus, volume $= 0.3225 \ \cancel{\text{mole NaOH}} \times \dfrac{L}{19.06 \ \cancel{\text{mole NaOH}}} = 0.0169 \ \text{L}$

Take 16.9 mL of the concentrated reagent and dilute to 1.50 L.

(f) $12 \ \text{ppm K}^+ = \dfrac{12 \ \text{mg K}^+}{\cancel{L}} \times 1.50 \ \cancel{L} = 18 \ \text{mg K}^+$

$$18 \ \cancel{mg} \times \frac{1 \ \cancel{g}}{1000 \ \cancel{mg}} \times \frac{\cancel{\text{mole K}^+}}{39.10 \ \cancel{g}} \times \frac{\cancel{\text{mole K}_4\text{Fe(CN)}_6}}{4 \ \cancel{\text{mole K}^+}} \times \frac{368.3 \ \text{g}}{\text{mole} \ \cancel{\text{K}_4\text{Fe(CN)}_6}}$$

$= 0.0424 \ \text{g} \ \text{K}_4\text{Fe(CN)}_6$

Dissolve 42.4 mg $\text{K}_4\text{Fe(CN)}_6$ in water and dilute to 1.50 L.

13-15. $\mathcal{M}_{Na_2CO_3} = 105.99 \ \dfrac{\text{g}}{\text{mole}}$

$$CO_3^{2-} + 2H^+ \rightleftharpoons H_2O + CO_2(g)$$

$$\frac{0.4723 \ \cancel{\text{g Na}_2\text{CO}_3} \times \dfrac{1 \ \cancel{\text{mole Na}_2\text{CO}_3}}{105.99 \ \cancel{\text{g Na}_2\text{CO}_3}} \times \dfrac{2 \ \cancel{\text{mole H}^+}}{\cancel{\text{mole Na}_2\text{CO}_3}} \times \dfrac{1 \ \cancel{\text{mole H}_2\text{SO}_4}}{2 \ \cancel{\text{mole H}^+}} \times \dfrac{1000 \ \text{mmol}}{\cancel{\text{mole H}_2\text{SO}_4}}}{34.78 \ \text{mL}}$$

$$= 0.1281 \text{ M}$$

13-17.　　$\dfrac{V_{HClO_4}}{V_{NaOH}} = \dfrac{26.93 \text{ mL HClO}_4}{25.00 \text{ mL NaOH}} = 1.0772 \dfrac{\text{mL HClO}_4}{\text{mL NaOH}}$

The volume of $HClO_4$ needed to titrate 0.4126 g of Na_2CO_3 is

$$40.00 \text{ mL HClO}_4 - 9.20 \text{ mL NaOH} \times \dfrac{1.0772 \text{ mL HClO}_4}{\text{mL NaOH}} = 30.09 \text{ mL}$$

Thus, $\dfrac{0.4126 \text{ g Na}_2CO_3}{30.09 \text{ mL HClO}_4} \times \dfrac{1 \text{ mmol Na}_2CO_3}{0.10588 \text{ g Na}_2CO_3} \times \dfrac{2 \text{ mmol HClO}_4}{\text{mmol Na}_2CO_3} = 0.2590 \text{ M HClO}_4$

and　　　$c_{NaOH} = c_{HClO_4} \times \dfrac{V_{HClO_4}}{V_{NaOH}}$

$$= \dfrac{0.2590 \text{ mmol HClO}_4}{\text{mL HClO}_4} \times \dfrac{1.0772 \text{ mL HClO}_4}{\text{mL NaOH}} \times \dfrac{1 \text{ mmol NaOH}}{\text{mmol HClO}_4} = 0.2790 \text{ M}$$

13-19.　　Each mole of KIO_3 consumes 6 moles of $S_2O_3^{2-}$

$$\dfrac{0.1142 \text{ g KIO}_3 \times \dfrac{1 \text{ mol KIO}_3}{214.001 \text{ g KIO}_3} \times \dfrac{1000 \text{ mmol Na}_2SO_3}{\text{mol Na}_2SO_3} \times \dfrac{6 \text{ mol Na}_2SO_3}{\text{mol KIO}_3}}{27.95 \text{ mL Na}_2SO_3}$$

$$= 0.1146 \text{ M Na}_2SO_3$$

13-21. No. mmol $Fe^{2+} = 25.00 \text{ mL} \times \dfrac{0.002517 \text{ mmol Cr}_2O_7^{2-}}{\text{mL}} \times \dfrac{6 \text{ mmol Fe}^{2+}}{\text{mmol Cr}_2O_7^{2-}} = 0.37755$

$$= \text{no. mmol analyte Fe}^{2+} + \text{no. mmol Fe}^{2+} \text{ back titrated}$$

No. mmol analyte $Fe^{2+} = 0.37755 - 8.53 \times 0.00949 \text{ M} = 0.2966$

$$\dfrac{0.2966 \text{ mmol Fe}}{100 \text{ mL}} \times \dfrac{0.055845 \text{ g}}{\text{mmol Fe}} \times \dfrac{1 \text{ mL}}{\text{g}} \times 10^6 \text{ ppm} = 165.6 \text{ ppm Fe}$$

13-23. $$\dfrac{37.31 \text{ mL Hg}^{2+} \times \dfrac{0.009372 \text{ mmol Hg}^{2+}}{\text{mL Hg}^{2+}} \times \dfrac{4 \text{ mmol}}{\text{mmol Hg}^{2+}} \times \dfrac{0.07612 \text{ g}}{\text{mmol}}}{1.455 \text{ g}} \times 100\%$$

$$= 7.317 \% \text{ (NH}_2)_2\text{CS}$$

13-25. Total amount KOH = 40.00 mL × 0.04672 mmol/mL = 1.8688 mmol

KOH reacting with H_2SO_4

$$= 3.41 \text{ mL H}_2\text{SO}_4 \times \dfrac{0.05042 \text{ mmol H}_2\text{SO}_4}{\text{mL H}_2\text{SO}_4} \times \dfrac{2 \text{ mmol KOH}}{\text{mmol H}_2\text{SO}_4} = 0.34386 \text{ mmol}$$

mass EtOAc = $(1.8688 - 0.34386)$ mmol KOH $\times \dfrac{1 \text{ mmol EtOAc}}{\text{mmol KOH}} \times \dfrac{0.08811 \text{ g}}{\text{mmol EtOAc}}$

$= 0.13436$ g in the 20.00-mL portion. In the entire 100.00-mL there are

5×0.13326 g or 0.6718 g.

13-27. (a) $\dfrac{0.3147 \text{ g Na}_2\text{C}_2\text{O}_4}{0.1340 \text{ g Na}_2\text{C}_2\text{O}_4 \text{ / mmol Na}_2\text{C}_2\text{O}_4} \times \dfrac{2 \text{ mmol KMnO}_4}{5 \text{ mmol Na}_2\text{C}_2\text{O}_4} = 0.9394 \text{ mmol KMnO}_4$

$$\dfrac{0.9394 \text{ mmol KMnO}_4}{31.67 \text{ mL}} = 0.02966 \text{ M KMnO}_4$$

(b) $MnO_4^- + 5Fe^{2+} + 8H^+ \rightleftharpoons Mn^{2+} + 5Fe^{3+} + 4H_2O$

26.75 mL $KMnO_4 \times 0.02966$ M = 0.7934 mmol $KMnO_4$. Each mmol $KMnO_4$ consumes

5 mmol Fe^{2+}. So

mmol $Fe^{2+} = 5 \times 0.7934 = 3.967$

$$\dfrac{3.967 \text{ mmol Fe}^{2+} \times \dfrac{1 \text{ mmol Fe}_2\text{O}_3}{2 \text{ mmol Fe}^{2+}} \times \dfrac{0.15969 \text{ g Fe}_2\text{O}_3}{\text{mmol Fe}_2\text{O}_3}}{0.6656 \text{ g}} \times 100\% = 47.59\%$$

13-29. (a) $c = \dfrac{7.48 \text{ g} \times \dfrac{1 \text{ mol}}{277.85 \text{ g}}}{2.000 \text{ L}} = 1.35 \times 10^{-2} \text{ M}$

(b) $[Mg^{2+}] = 1.35 \times 10^{-2} \text{ M}$

(c) There are 3 moles of Cl$^-$ for each mole of KCl•MgCl$_2$•6H$_2$O. Hence,

$[Cl^-] = 3 \times 1.346 \times 10^{-2} = 4.038 \times 10^{-2} \text{ M}$

(d) $\dfrac{7.48 \text{ g}}{2.00 \text{ L}} \times \dfrac{1 \text{ L}}{1000 \text{ mL}} \times 100\% = 0.374\% \text{ (w/v)}$

(e)

$\dfrac{1.346 \times 10^{-2} \text{ mmol } \cancel{KCl \cdot MgCl}_2}{\cancel{mL}} \times \dfrac{3 \text{ mmol Cl}^-}{\text{mmol } \cancel{KCl \cdot MgCl}_2} \times 25.00 \text{ } \cancel{mL} = 1.0095 \text{ mmol Cl}^-$

(f)

$\dfrac{1.346 \times 10^{-2} \text{ mmol KCl} \cdot \text{MgCl}_2}{\text{mL}} \times \dfrac{1 \text{ mmol K}^+}{\text{mmol KCl} \cdot \text{MgCl}_2} \times \dfrac{39.10 \text{ mg}}{\text{mmol K}^+} \times \dfrac{1000 \text{ mL}}{\text{L}} = \dfrac{526 \text{ mg K}^+}{\text{L}}$

$= 526 \text{ ppm K}^+$

$\dfrac{1.49 \times 10^{-3} \text{ mol}}{\text{L}} \times \dfrac{212.0 \text{ g}}{\text{mol}} \times \dfrac{1 \text{ L}}{1000 \text{ g}} \times 10^6 \text{ ppm} = 316 \text{ ppm}$

Chapter 14

14-1. The eye has limited sensitivity. To see the color change requires a roughly tenfold excess

of one or the other form of the indicator. This change corresponds to a pH range of the

indicator $pK_a \pm 1$ pH unit, a total range of 2 pH units.

14-3. **(a)** The initial pH of the NH_3 solution will be less than that for the solution containing

NaOH. With the first addition of titrant, the pH of the NH_3 solution will decrease rapidly

and then level off and become nearly constant throughout the middle part of the titration.

In contrast, additions of standard acid to the NaOH solution will cause the pH of the

NaOH solution to decrease gradually and nearly linearly until the equivalence point is

approached. The equivalence point pH for the NH_3 solution will be well below 7,

whereas for the NaOH solution it will be exactly 7.

(b) Beyond the equivalence point, the pH is determined by the excess titrant. Thus, the

curves become identical in this region.

14-5. The variables are temperature, ionic strength, and the presence of organic solvents and

colloidal particles.

14-6. The sharper end point will be observed with the solute having the larger K_b.

(a)   For NaOCl, $K_b = \dfrac{1.00 \times 10^{-14}}{3.0 \times 10^{-8}} = 3.3 \times 10^{-7}$

For hydroxylamine $K_b = \dfrac{1.00 \times 10^{-14}}{1.1 \times 10^{-6}} = 9.1 \times 10^{-9}$ Thus, NaOCl

(c)   For hydroxylamine $K_b = 9.1 \times 10^{-9}$ (part a)

For methyl amine, $K_b = \dfrac{1.00 \times 10^{-14}}{2.3 \times 10^{-11}} = 4.3 \times 10^{-4}$ Thus, methyl amine

14-7. The sharper end point will be observed with the solute having the larger K_a.

    (a)   For nitrous acid $K_a = 7.1 \times 10^{-4}$

              For iodic acid $K_a = 1.7 \times 10^{-1}$ Thus, iodic acid

    (c)   For hypochlorous acid $K_a = 3.0 \times 10^{-8}$

              For pyruvic acid $K_a = 3.2 \times 10^{-3}$ Thus, pyruvic acid

14-9. $InH^+ + H_2O \rightleftharpoons In + H_3O^+$ $\dfrac{[H_3O^+][In]}{[InH^+]} = K_a$

For methyl orange, $pK_a = 3.46$ (Table 14-1)

$K_a = \text{antilog}(-3.46) = 3.47 \times 10^{-4}$

$[InH^+]/[In] = 1.84$

Substituting these values into the equilibrium expression and rearranging gives

$$[H_3O^+] = 3.47 \times 10^{-4} \times 1.84 = 6.385 \times 10^{-4}$$

$$pH = -\log(6.385 \times 10^{-4}) = 3.19$$

14-11. (b)   At 50°C, $pK_w = -\log(5.47 \times 10^{-14}) = 13.26$

14-12. $pK_w = pH + pOH;\quad pOH = -\log[OH^-] = -\log(1.00 \times 10^{-2}) = 2.00$

    (b) $pH = 13.26 - 2.00 = 11.26$

14-13. $\dfrac{3.00 \text{ g HCl}}{100 \text{ g}} \times \dfrac{1.015 \text{ g}}{mL} \times \dfrac{1 \text{ mmol HCl}}{0.03646 \text{ g HCl}} = 0.835 \text{ M}$

$[H_3O^+] = 0.835 \text{ M};\quad pH = -\log 0.835 = 0.078$

14-15. The solution is so dilute that we must take into account the contribution of water to $[OH^-]$

which is equal to $[H_3O^+]$. Thus,

$$[OH^-] = 2.00 \times 10^{-8} + [H_3O^+] = 2.00 \times 10^{-8} + \dfrac{1.00 \times 10^{-14}}{[OH^-]}$$

$$[OH^-]^2 - 2.00 \times 10^{-8}[OH^-] - 1.00 \times 10^{-14} = 0$$

Solving the quadratic equation yields, $[OH^-] = 1.105 \times 10^{-7}$ M

$$pOH = -\log 1.105 \times 10^{-7} = 6.957; \quad pH = 14.00 - 6.957 = 7.04$$

14-17. amount of $Mg(OH)_2$ taken $= \dfrac{0.093 \ \text{g Mg(OH)}_2}{0.05832 \ \text{g Mg(OH)}_2 \ / \ \text{mmol}} = 1.595$ mmol

(a) $c_{HCl} = (75.0 \times 0.0500 - 1.595 \times 2)/75.0 = 7.467 \times 10^{-3}$ M

$[H_3O^+] = 7.467 \times 10^{-3}; \qquad pH = -\log(7.467 \times 10^{-3}) = 2.13$

(b) $c_{HCl} = 100.0 \times 0.0500 - 1.595 \times 2)/100.0 = 0.0181$ M

$pH = -\log(0.0181) = 1.74$

(c) $15.0 \times 0.050 = 0.750$ mmol HCl added. Solid $Mg(OH)_2$ remains and

$$[Mg^{2+}] = 0.750 \ \text{mmol HCl} \times \frac{1 \ \text{mmol Mg}^{2+}}{2 \ \text{mmol HCl}} \times \frac{1}{15.0 \ \text{mL}} = 0.0250 \ \text{M}$$

$$K_{sp} = 7.1 \times 10^{-12} = [Mg^{2+}][OH^-]^2$$

$$[OH^-] = \sqrt{\frac{7.1 \times 10^{-12}}{0.0250}} = 1.68 \times 10^{-5} \ \text{M}$$

$$pH = 14.00 - (-\log(1.68 \times 10^{-5})) = 9.22$$

(d) Since $Mg(OH)_2$ is fairly insoluble, the Mg^{2+} essentially all comes from the added

$MgCl_2$, and $[Mg^{2+}] = 0.0500$ M

$$[OH^-] = \sqrt{\frac{7.1 \times 10^{-12}}{0.0500}} = 1.19 \times 10^{-5} \ \text{M}$$

$$pH = 14.00 - (-\log(1.19 \times 10^{-5})) = 9.08$$

14-19. (a)　　$[H_3O^+] = 0.0500 \text{ M};$　$pH = -\log(0.0500) = 1.30$

(b)　　$\mu = \frac{1}{2}\{(0.0500)(+1)^2 + (0.0500)(-1)^2\} = 0.0500$

$\gamma_{H_3O^+} = 0.85 \text{ (Table 10-2)}$

$a_{H_3O^+} = 0.85 \times 0.0500 = 0.0425$

$pH = -\log(0.0425) = 1.37$

14-21.　$HOCl + H_2O \rightleftharpoons H_3O^+ + OCl^-$　　$K_a = \dfrac{[H_3O^+][OCl^-]}{[HOCl]} = 3.0 \times 10^{-8}$

$[H_3O^+] = [OCl^-]$ and $[HOCl] = c_{HOCl} - [H_3O^+]$

$[H_3O^+]^2/(c_{HOCl} - [H_3O^+]) = 3.0 \times 10^{-8}$

rearranging gives:　$[H_3O^+]^2 + 3 \times 10^{-8}[H_3O^+] - c_{HOCl} \times 3.0 \times 10^{-8} = 0$

	c_{HOCl}	$[H_3O^+]$	pH
(a)	0.100	5.476×10^{-5}	4.26
(b)	0.0100	1.731×10^{-5}	4.76
(c)	1.00×10^{-4}	1.717×10^{-6}	5.76

14-23.　$NH_3 + H_2O \rightleftharpoons NH_4^+ + OH^-$　　$K_b = \dfrac{1.00 \times 10^{-14}}{5.7 \times 10^{-10}} = 1.75 \times 10^{-5}$

$[NH_4^+] = [OH^-]$ and $[NH_3] = c_{NH_3} - [OH^-]$

$[OH^-]^2/(c_{NH_3} - [OH^-]) = 1.75 \times 10^{-5}$

rearranging gives:　$[OH^-]^2 + 1.75 \times 10^{-5}[OH^-] - c_{NH_3} \times 1.75 \times 10^{-5} = 0$

	c_{NH_3}	$[OH^-]$	pOH	pH
(a)	0.100	1.314×10^{-3}	2.88	11.12
(b)	0.0100	4.097×10^{-4}	3.39	10.62
(c)	1.00×10^{-4}	3.399×10^{-5}	4.47	9.53

14-25. $C_5H_{11}N + H_2O \rightleftharpoons C_5H_{11}NH^+ + OH^-$　　　$K_b = \dfrac{1.00 \times 10^{-14}}{7.5 \times 10^{-12}} = 1.333 \times 10^{-3}$

$[C_5H_{11}NH^+] = [OH^-]$ and $[C_5H_{11}N] = c_{C_5H_{11}N} - [OH^-]$

$[OH^-]^2/(c_{C_5H_{11}N} - [OH^-]) = 1.333 \times 10^{-3}$

rearranging gives: $[OH^-]^2 + 1.333 \times 10^{-3}[OH^-] - c_{C_5H_{11}N} \times 1.333 \times 10^{-3} = 0$

	$c_{C_5H_{11}N}$	$[OH^-]$	pOH	pH
(a)	0.100	1.090×10^{-2}	1.96	12.04
(b)	0.0100	3.045×10^{-3}	2.52	11.48
(c)	1.00×10^{-4}	9.345×10^{-5}	4.03	9.97

14-27. (a)

$c_{HA} = 36.5 \text{ g HA} \times \dfrac{1 \text{ mmol HA}}{0.090079 \text{ g HA}} \times \dfrac{1}{500 \text{ mL soln}} = 0.8104 \text{ M HA}$

$HL + H_2O \rightleftharpoons H_3O^+ + L^-$　　　$K_a = 1.38 \times 10^{-4}$

$[H_3O^+] = [L^-]$ and $[HL] = 0.8104 - [H_3O^+]$

$[H_3O^+]^2/(0.8104 - [H_3O^+]) = 1.38 \times 10^{-4}$

rearranging and solving the quadratic gives: $[H_3O^+] = 0.0105$ and pH = 1.98

(b)　　$c_{HA} = 0.8104 \times 25.0/250.0 = 0.08104 \text{ M HL}$

Proceeding as in part (a) we obtain: $[H_3O^+] = 3.28 \times 10^{-3}$ and pH = 2.48

(c)　　$c_{HA} = 0.08104 \times 10.0/1000.0 = 8.104 \times 10^{-4} \text{ M HL}$

Proceeding as in part (a) we obtain: $[H_3O^+] = 2.72 \times 10^{-4}$ and pH = 3.56

14-29. amount HFm taken = $20.00 \text{ mL} \times \dfrac{0.1750 \text{ mmol}}{\text{mL}} = 3.50$ mmol

   (a) $HFm + H_2O \rightleftharpoons H_3O^+ + Fm^-$ $K_a = 1.80 \times 10^{-4}$

$c_{HFm} = 3.50/45.0 = 7.78 \times 10^{-2}$ M

$[H_3O^+] = [Fm^-]$ and $[HFm] = 0.0778 - [H_3O^+]$

$[H_3O^+]^2/(0.0778 - [H_3O^+]) = 1.80 \times 10^{-4}$

rearranging and solving the quadratic gives: $[H_3O^+] = 3.65 \times 10^{-3}$ and pH = 2.44

   (b)   amount NaOH added = $25.0 \times 0.140 = 3.50$ mmol

Since all the formic acid has been neutralized, we are left with a solution of NaFm.

$Fm^- + H_2O \rightleftharpoons OH^- + HFm$ $K_b = 1.00 \times 10^{-14}/(1.80 \times 10^{-4}) = 5.56 \times 10^{-11}$

$c_{Fm^-} = 3.00/45.0 = 7.78 \times 10^{-2}$ M

$[OH^-] = [HFm]$ and $[Fm^-]$ $0.0778 - [OH^-]$

$[OH^-]^2/(0.0778 - [OH^-]) = 5.56 \times 10^{-11}$

rearranging and solving the quadratic gives: $[OH^-] = 2.08 \times 10^{-6}$ and pH = 8.32

   (c)   amount NaOH added = $25.0 \times 0.200 = 5.00$ mmol

therefore, we have an excess of NaOH; the pH is determined by the excess $[OH^-]$.

$[OH^-] = (5.00 - 3.50)/45.0 = 3.333 \times 10^{-2}$ M

pH = 14 − pOH = 12.52

   (d)   amount NaFm added = $25.0 \times 0.200 = 5.00$ mmol

$[HFm] = 3.50/45.0 = 0.0778$

$[Fm^-] = 5.00/45.00 = 0.1111$

$[H_3O^+] \times 0.1111/0.0778 = 1.80 \times 10^{-4}$

$[H_3O^+] = 1.260 \times 10^{-4}$ and pH $= 3.90$

14-31. (a) $NH_4^+ + H_2O \rightleftharpoons H_3O^+ + NH_3$ $K_a = 5.70 \times 10^{-10} = \dfrac{[H_3O^+][NH_3]}{[NH_4^+]}$

$[NH_3] = 0.0300$ M and $[NH_4^+] = 0.0500$ M

$[H_3O^+] = 5.70 \times 10^{-10} \times 0.0500/0.0300 = 9.50 \times 10^{-10}$ M

$[OH^-] = 1.00 \times 10^{-14}/9.50 \times 10^{-10} = 1.05 \times 10^{-5}$ M

pH $= -\log(9.50 \times 10^{-10}) = 9.02$

(b) $\mu = \frac{1}{2}\{(0.0500)(+1)^2 + (0.0500)(-1)^2\} = 0.0500$

From Table 10-2 $\gamma_{NH_4^+} = 0.80$ and $\gamma_{NH_3} = 1.0$

$$a_{H_3O^+} = \frac{K_a \gamma_{NH_4^+}[NH_4^+]}{\gamma_{NH_3}[NH_3]} = \frac{5.70 \times 10^{-5} \times 0.80 \times 0.0500}{1.00 \times 0.0300} = 7.60 \times 10^{-10}$$

pH $= -\log(7.60 \times 10^{-10}) = 9.12$

14-33. In each of the parts of this problem, we are dealing with a weak base B and its conjugate

acid BHCl or $(BH)_2SO_4$. The pH determining equilibrium can then be written as

$BH^+ + H_2O \rightleftharpoons H_3O^+ + B$

The equilibrium concentration of BH^+ and B are given by

$[BH^+] = c_{BHCl} + [OH^-] - [H_3O^+]$ (1)

$[B] = c_B - [OH^-] + [H_3O^+]$ (2)

In many cases $[OH^-]$ and $[H_3O^+]$ will be much smaller than c_B and c_{BHCl} and $[BH^+] \approx$

c_{BHCl} and $[B] \approx c_B$ so that

91

$$[H_3O^+] = K_a \times \frac{c_{BHCl}}{c_B} \qquad (3)$$

(a) Amount NH_4^+ = 3.30 ~~g $(NH_4)_2SO_4$~~ $\times \dfrac{1 \text{ mmol } \overline{(NH_4)_2SO_4}}{0.13214 \text{ g } \overline{(NH_4)_2SO_4}} \times \dfrac{2 \text{ mmol } NH_4^+}{\text{mmol } \overline{(NH_4)_2SO_4}}$

$$= 49.95 \text{ mmol}$$

Amount NaOH = 125.0 mL $\times$ 0.1011 mmol/mL = 12.64 mmol

$$c_{NH_3} = 12.64 \text{ ~~mmol NaOH~~} \times \frac{1 \text{ mmol } NH_3}{\text{~~mmol NaOH~~}} \times \frac{1}{500.0 \text{ mL}} = 2.528 \times 10^{-2} \text{ M}$$

$$c_{NH_4^+} = (49.95 - 12.64) \text{ mmol } NH_4^+ \times \frac{1}{500.0 \text{ mL}} = 7.462 \times 10^{-2} \text{ M}$$

Substituting these relationships in equation (3) gives

$$[H_3O^+] = K_a \times \frac{c_{BHCl}}{c_B} = 5.70 \times 10^{-10} \times 7.462 \times 10^{-2} /(2.528 \times 10^{-2}) = 1.682 \times 10^{-9} \text{ M}$$

$$[OH^-] = 1.00 \times 10^{-14}/1.682 \times 10^{-9} = 5.95 \times 10^{-6} \text{ M}$$

Note, $[H_3O^+]$ and $[OH^-]$ are small compared to c_{NH_3} and $c_{NH_4^+}$ so our assumption is valid.

$$pH = -\log(1.682 \times 10^{-9}) = 8.77$$

(b)   Substituting into equation (3) gives

$$[H_3O^+] = 7.5 \times 10^{-12} \times 0.010/0.120 = 6.25 \times 10^{-13} \text{ M}$$

$$[OH^-] = 1.00 \times 10^{-14}/6.25 \times 10^{-13} = 1.60 \times 10^{-2} \text{ M}$$

Again our assumption is valid and

$$pH = -\log(6.25 \times 10^{-13}) = 12.20$$

(c) c_B = 0.050 M and c_{BHCl} = 0.167 M

$$[H_3O^+] = 2.31 \times 10^{-11} \times 0.167/0.050 = 7.715 \times 10^{-11} \text{ M}$$

$$[OH^-] = 1.00 \times 10^{-14}/7.715 \times 10^{-11} = 1.30 \times 10^{-4} \text{ M}$$

92

The assumption is valid and

$$pH = -\log(7.715 \times 10^{-11}) = 10.11$$

(d)   Original amount B $= 2.32 \not{gB} \times \dfrac{1 \text{ mmol}}{0.09313 \not{gB}} = 24.91 \text{ mmol} = 24.91 \text{ mmol}$

Amount HCl $= 100 \text{ mL} \times 0.0200 \text{ mmol/mL} = 2.00 \text{ mmol}$

$c_B = (24.91 - 2.00)/250.0 = 9.164 \times 10^{-2} \text{ M}$

$c_{BH^+} = 2.00/250.0 = 8.00 \times 10^{-3} \text{ M}$

$[H_3O^+] = 2.51 \times 10^{-5} \times 8.00 \times 10^{-3}/(9.164 \times 10^{-2}) = 2.191 \times 10^{-6} \text{ M}$

$[OH^-] = 1.00 \times 10^{-14}/2.191 \times 10^{-6} = 4.56 \times 10^{-9} \text{ M}$

Our assumptions are valid, so

$$pH = -\log(2.191 \times 10^{-6}) = 5.66$$

14-34. (a)   0.00

(c)   pH diluted solution $= 14.000 - [-\log(0.00500)] = 11.699$

pH undiluted solution $= 14.000 - [-\log(0.0500)] = 12.699$

$\Delta pH = 11.699 - 12.699 = -1.000$

(e) $OAc^- + H_2O \rightleftharpoons HOAc + OH$

$$K_b = \dfrac{[HOAc][OH^-]}{[OAc^-]} = \dfrac{1.00 \times 10^{-14}}{1.75 \times 10^{-5}} = 5.71 \times 10^{-10}$$

Here we can use an approximation because K_b is very small. For the undiluted

sample:

$$\dfrac{[OH^-]^2}{0.0500} = 5.71 \times 10^{-10}$$

$$[OH^-] = (5.71 \times 10^{-10} \times 0.0500)^{1/2} = 5.343 \times 10^{-6} \text{ M}$$

93

$$pH = 14.00 - [-\log(5.343 \times 10^{-6})] = 8.728$$

For the diluted sample

$$[OH^-] = (5.71 \times 10^{-10} \times 0.00500)^{1/2} = 1.690 \times 10^{-6} \text{ M}$$

$$pH = 14.00 - [-\log(1.690 \times 10^{-6})] = 8.228$$

$$\Delta pH = 8.228 - 8.728 = -0.500$$

(g) Proceeding as in part (f) a 10-fold dilution of this solution results in a pH change that is less than 1 in the third decimal place. Thus for all practical purposes,

$$\Delta pH = 0.000$$

Note a more concentrated buffer compared to part (f) gives an even smaller pH change.

14-35. (a) After addition of acid, $[H_3O^+] = 1$ mmol/100 mL $= 0.0100$ M and pH $= 2.00$

Since original pH $= 7.00$

$$\Delta pH = 2.00 - 7.00 = -5.00$$

(b) After addition of acid

$$c_{HCl} = (100 \times 0.0500 + 1.00)/100 = 0.0600 \text{ M}$$

$$\Delta pH = -\log(0.0600) - [-\log(0.0500)] = 1.222 - 1.301 = -0.079$$

(c) After addition of acid,

$$c_{NaOH} = (100 \times 0.0500 - 1.00)/100 = 0.0400 \text{ M}$$

$$[OH^-] = 0.0400 \text{ M} \text{ and } pH = 14.00 - [-\log(0.0400)] = 12.602$$

From Problem 14-34 (c), original pH $= 12.699$

$$\Delta pH = -0.097$$

(d)   From Solution 14-34 (d), original pH = 3.033

Upon adding 1 mmol HCl to the 0.0500 M HOAc, we produce a mixture that is

0.0500 M in HOAc and 1.00/100 = 0.0100 M in HCl. The pH of this solution is

approximately that of a 0.0100 M HCl solution, or 2.00. Thus

ΔpH = 2.000 – 3.033 = –1.033

(If the contribution of the dissociation of HOAc to the pH is taken into account, a

pH of 1.996 is obtained and ΔpH = –1.037 is obtained.)

(e)   From Solution 14-34 (e), original pH = 8.728

Upon adding 1.00 mmol HCl we form a buffer having the composition

c_{HOAc} = 1.00/100 = 0.0100

c_{NaOAc} = (0.0500 × 100 – 1.00)/100 = 0.0400

$[H_3O^+]$ = 1.75 × 10^{-5} × 0.0100/0.0400 = 4.575 × 10^{-6} M

pH = –log(4.575 × 10^{-6}) = 5.359

ΔpH = 5.359 – 8.728 = –3.369

(f)   From Solution 14-34 (f), original pH = 4.757

With the addition of 1.00 mmol of HCl we have a buffer whose concentrations are

c_{HOAc} = 0.0500 + 1.00/100 = 0.0600 M

c_{NaOAc} = 0.0500 – 1.00/100 = 0.0400 M

Proceeding as in part (e), we obtain

$[H_3O^+]$ = 2.625 × 10^{-5} M and pH = 4.581

ΔpH = 4.581 – 4.757 = –0.176

Note again the very small pH change as compared to unbuffered solutions.

(g)   For the original solution

$$[H_3O^+] = 1.75 \times 10^{-5} \times 0.500/0.500 = 1.75 \times 10^{-5} \text{ M}$$

$$pH = -\log(1.75 \times 10^{-5}) = 4.757$$

After addition of 1.00 mmol HCl

$$c_{HOAc} = 0.500 + 1.00/100 = 0.510 \text{ M}$$

$$c_{NaOAc} = 0.500 - 1.00/100 = 0.490 \text{ M}$$

Proceeding as in part (e), we obtain

$$[H_3O^+] = 1.75 \times 10^{-5} \times 0.510/0.490 = 1.821 \times 10^{-5} \text{ M}$$

$$pH = -\log(1.821 \times 10^{-5}) = 4.740$$

$$\Delta pH = 4.740 - 4.757 = -0.017$$

Note that the more concentrated buffer is even more effective in resisting pH changes.

14-37. For lactic acid, $K_a = 1.38 \times 10^{-4} = [H_3O^+][L^-]/[HL]$

Throughout this problem we will base calculations on Equations 9-25 and 9-26

$$[L^-] = c_{NaL} + [H_3O^+] - [OH^-] \approx c_{NaL} + [H_3O^+]$$

$$[HL] = c_{HL} - [H_3O^+] - [OH^-] \approx c_{HL} - [H_3O^+]$$

$$\frac{[H_3O^+]\left(c_{NaL} + [H_3O^+]\right)}{c_{HL} - [H_3O^+]} = 1.38 \times 10^{-4}$$

This equation rearranges to

$$[H_3O^+]^2 + (1.38 \times 10^{-4} + 0.0800)[H_3O^+] - 1.38 \times 10^{-4} \times c_{HL} = 0$$

(b)   Before addition of acid

$$[H_3O^+]^2 + (1.38 \times 10^{-4} + 0.0200)[H_3O^+] - 1.38 \times 10^{-4} \times 0.0800 = 0$$

$$[H_3O^+] = 5.341 \times 10^{-5} \text{ and } pH = 3.272$$

After adding acid

$$c_{HL} = (100 \times 0.0800 + 0.500)/100 = 0.0850 \text{ M}$$

$$c_{NaL} = (100 \times 0.0200 - 0.500)/100 = 0.0150 \text{ M}$$

$$[H_3O^+]^2 + (1.38 \times 10^{-4} + 0.0150)[H_3O^+] - 1.38 \times 10^{-4} \times 0.0850 = 0$$

$$[H_3O^+] = 7.388 \times 10^{-4} \text{ and } pH = 3.131$$

$$\Delta pH = 3.131 - 3.272 = -0.141$$

14-39. The end point will occur when 25.00 mL of titrant have been added. Let us calculate pH when 24.95 and 25.05 mL of reagent have been added.

$$c_{A^-} \approx \frac{\text{amount KOH added}}{\text{total volume soln}} = \frac{24.95 \times 0.1000 \text{ mmol KOH}}{74.95 \text{ mL soln}} = \frac{2.495}{74.95} = 0.03329 \text{ M}$$

$$c_{HA} \approx [HA] = \frac{\text{original amount HA} - \text{amount KOH added}}{\text{total volume soln}}$$

$$= \frac{(50.00 \times 0.0500 - 24.95 \times 0.1000) \text{ mmol HA}}{74.95 \text{ mL soln}}$$

$$= \frac{2.500 - 2.495}{74.95} = \frac{0.005}{74.95} = 6.67 \times 10^{-5} \text{ M}$$

Substituting into Equation 9-29

$$[H_3O^+] = K_a \frac{c_{HA}}{c_{A^-}} = \frac{1.80 \times 10^{-4} \times 6.67 \times 10^{-5}}{0.03329} = 3.607 \times 10^{-7} \text{ M}$$

$$pH = -\log(3.607 \times 10^{-7}) = 6.44$$

At 25.05 mL KOH

$$c_{KOH} = [OH^-] = \frac{\text{amount KOH added} - \text{initial amount HA}}{\text{total volume soln}}$$

$$= \frac{25.05 \times 0.1000 - 50.00 \times 0.05000}{75.05 \text{ mL soln}} = 6.66 \times 10^{-5} \text{ M}$$

97

$$pH = 14.00 - [-\log(6.66 \times 10^{-5})] = 9.82$$

Thus, the indicator should change color in the range of pH 6.5 to 9.8. Cresol

purple (range 7.6 to 9.2, Table 14-1) would be quite suitable.

Problems 14-41 through 14-43. We will set up spreadsheets that will solve a quadratic

equation to determine $[H_3O^+]$ or $[OH^-]$, as needed. While approximate solutions are appropriate

for many of the calculations, the approach taken represents a more general solution and is

somewhat easier to incorporate in a spreadsheet. As an example consider the titration of a weak

acid with a strong base. Here c_a and V_i represent initial concentration and initial volume.

Before the equivalence point: $\qquad$ $[HA] = \dfrac{(c_{i\,HA}V_{i\,HA} - c_{i\,NaOH}V_{NaOH})}{(V_{i\,HA} + V_{NaOH})} - [H_3O^+]$

and $\qquad\qquad\qquad\qquad\qquad$ $[OH^-] = \dfrac{(c_{i\,NaOH}V_{NaOH} - c_{i\,HA}V_{i\,HA})}{(V_{i\,HA} + V_{NaOH})} + [HA]$

Substituting these expressions into the equilibrium expression for [HA] and rearranging gives

$$[H_3O^+]^2 + \left(\frac{(c_{i\,NaOH}V_{NaOH})}{(V_{i\,HA} + V_{NaOH})} + K_a\right)[H_3O^+] - \frac{K_a(c_{i\,HA}V_{i\,HA} - c_{i\,NaOH}V_{NaOH})}{(V_{i\,HA} + V_{NaOH})} = 0$$

From which $[H_3O^+]$ is directly determined.

At and after the equivalence point: $\qquad$ $[A^-] = \dfrac{(c_{i\,HA}V_{HA})}{(V_{i\,HA} + V_{NaOH})} - [HA]$

$$[OH^-] = \dfrac{(c_{i\,NaOH}V_{NaOH} - c_{i\,HA}V_{i\,HA})}{(V_{i\,HA} + V_{NaOH})} + [HA]$$

Substituting these expressions into the equilibrium expression for [A⁻] and rearranging gives

$$[HA]^2 + \left(\frac{(c_{i\,NaOH} V_{NaOH} - c_{i\,HA} V_{i\,HA})}{(V_{i\,HA} + V_{NaOH})} + \frac{K_w}{K_a} \right)[HA] - \frac{K_w (c_{i\,HA} V_{HA})}{K_a (V_{i\,HA} + V_{NaOH})} = 0$$

From which [HA] can be determined and [OH⁻] and [H₃O⁺] subsequently calculated. A similar

approach is taken for the titration of a weak base with a strong acid.

14-41.

	A	B	C	D	E	F	G
1	**Pb 14-41(a)**						
2	V_i HNO$_2$	50.00					
3	c_i HNO$_2$	0.1000					
4	c_i NaOH	0.1000					
5	K_a for HNO$_2$	7.10E-04					
6	K_w	1.00E-14					
7	V_{ep}	50.00					
8							
9	V_{NaOH}, mL	b in quadratic	c in quadratic	[HNO$_2$]	[OH$^-$]	[H$_3$O$^+$]	pH
10	0.00	7.1000E-04	-7.1000E-05			8.0786E-03	2.0927
11	5.00	9.8009E-03	-5.8091E-05			4.1607E-03	2.3808
12	15.00	2.3787E-02	-3.8231E-05			1.5112E-03	2.8207
13	25.00	3.4043E-02	-2.3667E-05			6.8155E-04	3.1665
14	40.00	4.5154E-02	-7.8889E-06			1.7404E-04	3.7594
15	45.00	4.8078E-02	-3.7368E-06			7.7599E-05	4.1101
16	49.00	5.0205E-02	-7.1717E-07			1.4281E-05	4.8452
17	50.00	1.4085E-11	-7.0423E-13	8.39174E-07	8.3917E-07	1.1916E-08	7.9239
18	51.00	9.9010E-04	-6.9725E-13	7.04225E-10	9.9010E-04	1.0100E-11	10.9957
19	55.00	4.7619E-03	-6.7069E-13	1.40845E-10	4.7619E-03	2.1000E-12	11.6778
20	60.00	9.0909E-03	-6.4020E-13	7.04225E-11	9.0909E-03	1.1000E-12	11.9586
21	**Spreadsheet Documentation**						
22	Cell B7=B2*B3/B4						
23	Cell B10=$B$4*A10/($B$2+A10)+$B$5						
24	Cell C10=-$B$5*($B$3*$B$2-$B$4*A10)/($B$2+A10)						
25	Cell F10=(-B10+SQRT(B10^2-4*C10))/2						
26	Cell G10=-LOG(F10)						
27	Cell B17=($B$4*A17-$B$3*$B$2)/($B$2+A17)+$B$6/$B$5						
28	Cell C17=-$B$6*$B$3*$B$2/($B$5*($B$2+A17))						
29	Cell D17=(-B17+SQRT(B17^2-4*C17))/2						
30	Cell E17=($B$4*A17-$B$3*$B$2)/($B$2+A17)+D17						
31	Cell F17=$B$6/E17						

	A	B	C	D	E	F	G
1	Pb 14-41(c)						
2	V_i HL	50.00					
3	c_i HL	0.1000					
4	c_i NaOH	0.1000					
5	K_a for HL	1.38E-04					
6	K_w	1.00E-14					
7	V_{ep}	50.00					
8							
9	V_{NaOH}, mL	b in quadratic	c in quadratic	[HL]	[OH⁻]	[H$_3$O⁺]	pH
10	0.00	1.3800E-04	-1.3800E-05			3.6465E-03	2.4381
11	5.00	9.2289E-03	-1.12909E-05			1.0938E-03	2.9611
12	15.00	2.3215E-02	-7.43077E-06			3.1579E-04	3.5006
13	25.00	3.3471E-02	-0.0000046			1.3687E-04	3.8637
14	40.00	4.4582E-02	-1.53333E-06			3.4367E-05	4.4639
15	45.00	4.7506E-02	-7.26316E-07			1.5284E-05	4.8158
16	49.00	4.9633E-02	-1.39394E-07			2.8083E-06	5.5516
17	50.00	7.2464E-11	-3.62319E-12	1.9E-06	1.9034E-06	5.2537E-09	8.2795
18	51.00	9.9010E-04	-3.5873E-12	3.62E-09	9.9010E-04	1.0100E-11	10.9957
19	55.00	4.7619E-03	-3.4507E-12	7.25E-10	4.7619E-03	2.1000E-12	11.6778
20	60.00	9.0909E-03	-3.2938E-12	3.62E-10	9.0909E-03	1.1000E-12	11.9586
21	**Spreadsheet Documentation**						
22	Cell B7=B2*B3/B4						
23	Cell B10=$B$4*A10/($B$2+A10)+$B$5						
24	Cell C10=-$B$5*($B$3*$B$2-$B$4*A10)/($B$2+A10)						
25	Cell F10=(-B10+SQRT(B10^2-4*C10))/2						
26	Cell G10=-LOG(F10)						
27	Cell B17=($B$4*A17-$B$3*$B$2)/($B$2+A17)+$B$6/$B$5						
28	Cell C17=-$B$6*$B$3*$B$2/($B$5*($B$2+A17))						
29	Cell D17=(-B17+SQRT(B17^2-4*C17))/2						
30	Cell E17=($B$4*A17-$B$3*$B$2)/($B$2+A17)+D17						
31	Cell F17=$B$6/E17						

14-43 (a) This titration of a weak acid with strong base follows the same basic spreadsheet as Pb 14-41 with the concentrations changed. A Scatter plot of pH vs. volume of NaOH is produced from the data.

	A	B	C	D	E	F	G
1	Pb 14-43(a)						
2	V_i ClCH$_2$COOH	50.00					
3	c_i ClCH$_2$COOH	0.0100					
4	c_i NaOH	0.0100					
5	K_a for ClCH$_2$COOH	1.36E-03					
6	K_w	1.00E-14					
7	V_{ep}	50.00					
8							
9	V_{NaOH}, mL	b in quadratic	c in quadratic	[HA]	[OH$^-$]	[H$_3$O$^+$]	pH
10	0.00	1.3600E-03	-1.3600E-05			3.0700E-03	2.5129
11	5.00	2.2691E-03	-1.11273E-05			2.3889E-03	2.6218
12	15.00	3.6677E-03	-7.32308E-06			1.4351E-03	2.8431
13	25.00	4.6933E-03	-4.53333E-06			8.2196E-04	3.0852
14	40.00	5.8044E-03	-1.51111E-06			2.4960E-04	3.6027
15	45.00	6.0968E-03	-7.15789E-07			1.1523E-04	3.9385
16	49.00	6.3095E-03	-1.37374E-07			2.1698E-05	4.6636
17	50.00	7.3529E-12	-3.67647E-14	1.92E-07	1.9174E-07	5.2155E-08	7.2827
18	51.00	9.9010E-05	-3.6401E-14	3.68E-10	9.9010E-05	1.0100E-10	9.9957
19	55.00	4.7619E-04	-3.5014E-14	7.35E-11	4.7619E-04	2.1000E-11	10.6778
20	60.00	9.0909E-04	-3.3422E-14	3.68E-11	9.0909E-04	1.1000E-11	10.9586
21	**Spreadsheet Documentation**						
22	Same as Pb 14-41a						

(c)

	A	B	C	D	E	F	G
1	**Pb 14-43(c)**						
2	V_i HOCl	50.00					
3	c_i HOCl	0.1000					
4	c_i NaOH	0.1000					
5	K_a for HOCl	3.00E-08					
6	K_w	1.00E-14					
7	V_{ep}	50.00					
8							
9	V_{NaOH}, mL	b in quadratic	c in quadratic	[HA]	[OH⁻]	[H₃O⁺]	pH
10	0.00	3.0000E-08	-3.0000E-09			5.4757E-05	4.2616
11	5.00	9.0909E-03	-2.4545E-09			2.6999E-07	6.5687
12	15.00	2.3077E-02	-1.6154E-09			7.0000E-08	7.1549
13	25.00	3.3333E-02	-1.0000E-09			3.0000E-08	7.5229
14	40.00	4.4444E-02	-3.3333E-10			7.5000E-09	8.1249
15	45.00	4.7368E-02	-1.5789E-10			3.3333E-09	8.4771
16	49.00	4.9495E-02	-3.0303E-11			6.1224E-10	9.2131
17	50.00	3.3333E-07	-1.6667E-08	0.000129	1.2893E-04	7.7560E-11	10.1104
18	51.00	9.9043E-04	-1.6502E-08	1.64E-05	1.0065E-03	9.9355E-12	11.0028
19	55.00	4.7622E-03	-1.5873E-08	3.33E-06	4.7652E-03	2.0985E-12	11.6781
20	60.00	9.0912E-03	-1.5152E-08	1.67E-06	9.0926E-03	1.0998E-12	11.9587
21	**Spreadsheet Documentation**						
22	Same as Pb 14-41a						

14-44. Here, we make use of Equations 9-36 And 9-37:

$$\alpha_0 = \frac{[H_3O^+]}{[H_3O^+] + K_a} \qquad \alpha_1 = \frac{K_a}{[H_3O^+] + K_a}$$

	A	B	C	D	E	F	G
1		Species	pH	[H$_3$O$^+$]	K$_a$	α_0	α_1
2	(a)	Acetic acid	5.320	4.7863E-06	1.75E-05	0.215	0.785
3	(b)	Picric acid	1.250	5.6234E-02	4.30E-01	0.116	0.884
4	(c)	HOCl	7.000	1.0000E-07	3.00E-08	0.769	0.231
5	(d)	HONH$_3$$^+$	5.120	7.5858E-06	1.10E-06	0.873	0.127
6	(e)	Piperdine	10.080	8.3176E-11	7.50E-12	0.917	0.083
7							
8	**Spreadsheet Documentation**						
9	Cell D2=10^(-C2)						
10	Cell F2=D2/(D2+E2)						
11	Cell G2=E2/(D2+E2						

14-45. $[H_3O^+] = 3.38 \times 10^{-12}$ M. For $CH_3NH_3^+$, Equation 9-37 takes the form,

$$\alpha_1 = \frac{[CH_3NH_2]}{c_T} = \frac{K_a}{[H_3O^+] + K_a} = \frac{2.3 \times 10^{-11}}{3.38 \times 10^{-12} + 2.3 \times 10^{-11}} = 0.872$$

$$[CH_3NH_2] = 0.872 \times 0.120 = 0.105 \text{ M}$$

14-47. For lactic acid, $K_a = 1.38 \times 10^{-4}$

$$\alpha_0 = \frac{[H_3O^+]}{K_a + [H_3O^+]} = \frac{[H_3O^+]}{1.38 \times 10^{-4} + [H_3O^+]}$$

$$\alpha_0 = 0.640 = \frac{[HA]}{c_T} = \frac{[HA}{0.120}$$

$$[HA] = 0.640 \times 0.120 = 0.0768 \text{ M}$$

$$\alpha_1 = 1.000 - 0.640 = 0.360$$

$$[A^-] = \alpha_1 \times 0.120 = (1.000 - 0.640) \times 0.120 = 0.0432 \text{ M}$$

$$[H_3O^+] = K_a \times c_{HA}/c_{A-} = 1.38 \times 10^{-4} \times 0.640/(1 - 0.640) = 2.453 \times 10^{-4} \text{ M}$$

$$pH = -\log(2.453 \times 10^{-4}) = 3.61$$

The remaining entries in the table are obtained in a similar manner. Bolded entries are the missing data points.

Acid	c_T	pH	[HA]	[A$^-$]	α_0	α_1
Lactic	0.120	**3.61**	**0.0768**	**0.0432**	0.640	**0.360**
Butanoic	**0.162**	5.00	0.644	**0.0979**	0.397	**0.604**
Sulfamic	0.250	1.20	**0.095**	**0.155**	**0.380**	**0.620**

Chapter 15

15-1. Not only is NaHA a proton donor, it is also the conjugate base of the parent acid H_2A.

$$HA^- + H_2O \rightleftharpoons H_3O^+ + A^{2-}$$

$$HA^- + H_2O \rightleftharpoons H_2A + OH^-$$

Solutions of acid salts can be acidic or alkaline, depending on which of the above equilibria predominates. In order to calculate the pH of solutions of this type, it is necessary to take both equilibria into account.

15-4. The species HPO_4^{2-} is such a weak acid ($K_{a3} = 4.5 \times 10^{-13}$) that the change in pH in the vicinity of the third equivalence point is too small to be observable.

15-5. **(a)** $NH_4^+ + H_2O \rightleftharpoons NH_3 + H_3O^+$ $K_a = 5.70 \times 10^{-10}$

$$OAc^- + H_2O \rightleftharpoons HOAc + OH^-   K_b = \frac{K_w}{K_a} = \frac{1.00 \times 10^{-14}}{1.75 \times 10^{-5}} = 5.71 \times 10^{-10}$$

Since the K's are essentially identical, the solution should be approximately neutral

(c)   Neither K^+ nor NO_3^- reacts with H_2O.   Solution will be neutral

(e) $C_2O_4^{2-} + H_2O \rightleftharpoons HC_2O_4^- + OH^-$ $K_b = \dfrac{1.00 \times 10^{-14}}{5.42 \times 10^{-5}} = 1.84 \times 10^{-10}$

Solution will be basic

(g) $H_2PO_4^- + H_2O \rightleftharpoons HPO_4^{2-} + H_3O^+$ $K_{a2} = 6.32 \times 10^{-8}$

$$H_2PO_4^- + H_2O \rightleftharpoons H_3PO_4 + OH^-   K_{b3} = \frac{1.00 \times 10^{-14}}{7.11 \times 10^{-3}} = 1.4 \times 10^{-12}$$

Solution will be acidic

15-6. We can approximate the $[H_3O^+]$ at the first equivalence point by Equation 15-16. Thus,

$$[H_3O^+] = \sqrt{5.8 \times 10^{-3} \times 1.1 \times 10^{-7}} = 2.53 \times 10^{-5}$$

$$pH = -\log(2.53 \times 10^{-5}) = 4.60$$

Bromocresol green would be a satisfactory indicator.

15-8. Curve A in figure 15-4 is the titration curve for H_3PO_4. Note that one end point occurs at about pH 4.5 and a second at about pH 9.5. Thus, H_3PO_4 would be determined by titration with bromocresol green as an indicator (pH 3.8 to 5.4). A titration to the second end point with phenolphthalein would give the number of millimoles of NaH_2PO_4 plus twice the number of millimoles of H_3PO_4. Thus, the concentration of NaH_2PO_4 is obtained from the difference in volume for the two titrations.

15-9. **(a)** To obtain the approximate equivalence point pH, we employ Equation 15-16

$$[H_3O^+] = \sqrt{K_{a1}K_{a2}} = \sqrt{4.2 \times 10^{-7} \times 4.69 \times 10^{-11}} = 4.4 \times 10^{-9}$$

$$pH = 8.4$$

Cresol purple (7.6 to 9.2) would be suitable.

(c)   As in part (b)

$$[OH^-] = (0.05 \times 1.00 \times 10^{-14}/4.31 \times 10^{-5})^{1/2} = 3.41 \times 10^{-6} \text{ M}$$

$$pH = 14.00 - [-\log(3.41 \times 10^{-6})] = 8.53$$

Cresol purple (7.6 to 9.2)

(e) $NH_3C_2H_4NH_3^{2+} + H_2O \rightleftharpoons NH_3C_2H_4NH_2^+ + H_3O^+$ $K_{a1} = 1.42 \times 10^{-7}$

$$[H_3O^+] = (0.05 \times 1.42 \times 10^{-7})^{1/2} = 8.43 \times 10^{-5} \text{ M}$$

$$pH = -\log(8.43 \times 10^{-5}) = 4.07$$

Bromocresol green (3.8 to 5.4)

(g)　　Proceeding as in part (b) we obtain pH = 9.94

　　　　Phenolphthalein (8.5 to 10.0)

15-10. (a)　　$H_3PO_4 + H_2O \rightleftharpoons H_3O^+ + H_2PO_4^-$　　$K_{a1} = 7.11 \times 10^{-3}$

$$\frac{[H_3O^+][H_2PO_4^-]}{[H_3PO_4]} = \frac{[H_3O^+]^2}{0.040 - [H_3O^+]} = 7.11 \times 10^{-3}$$

$$[H_3O^+]^2 + 7.11 \times 10^{-3}[H_3O^+] - 0.040 \times 7.11 \times 10^{-3} = 0$$

Solving by the quadratic formula or by successive approximations, gives

$[H_3O^+] = 1.37 \times 10^{-2}$ M　　　　pH $= -\log(1.37 \times 10^{-2}) = 1.86$

(c)　　pH = 1.64

(e)　　pH = 4.21

15-11. Throughout this problem, we will use Equation 15-15 or one of its simplificationws.

(a)　　$[H_3O^+] = \sqrt{\dfrac{0.0400 \times 6.32 \times 10^{-8}}{1 + 0.0400/(7.11 \times 10^{-3})}} = 1.95 \times 10^{-5}$

(c)　　pH = 4.28

(e)　　pH = 9.80

15-12. (a)　　$PO_4^{3-} + H_2O \rightleftharpoons HPO_4^{2-} + OH^-$　　$K_b = \dfrac{K_w}{K_{a3}} = \dfrac{1.00 \times 10^{-14}}{4.5 \times 10^{-13}} = 2.2. \times 10^{-2}$

$$\frac{[OH^-]^2}{0.040 - [OH^-]} = 2.22 \times 10^{-2}$$

$$[OH^-]^2 + 2.22 \times 10^{-2}[OH^-] - 8.88 \times 10^{-4} = 0$$

Solving gives　　　　$[OH^-] = 2.07 \times 10^{-2}$ M

　　　　pH $= 14.00 - [-\log(2.07 \times 10^{-2})] = 12.32$

(c)　　Proceeding as in part (b), we obtain pH = 9.70.

(e)   Proceeding as in part (a), gives pH = 12.58

15-14. (a)   Proceeding as in Problem 15-12(a), $[H_3O^+] = 3.77 \times 10^{-3}$ M and pH = 2.42

(b)   Proceeding as in 15-12(b), $[H_3O^+] = 3.10 \times 10^{-8}$ M and pH = 7.51

(c) $HOC_2H_4NH_3^+ + H_2O \rightleftharpoons HOC_2H_4NH_2 + H_3O^+ \quad K_a = 3.18 \times 10^{-10}$

Proceeding as in 15-12(b) we obtain $[H_3O^+] = 3.73 \times 10^{-10}$ M and pH = 9.43

(d) $H_2C_2O_4 + C_2O_4^{2-} \rightarrow 2HC_2O_4^-$

For each milliliter of solution, 0.0240 mmol H_2HPO_4 reacts with 0.0240 mmol

$C_2O_4^{2-}$ to give 0.0480 mmol $HC_2O_4^-$ and to leave 0.0120 mmol $C_2O_4^{2-}$. Thus, we

have a buffer that is 0.0480 M in $HC_2O_4^-$ and 0.0120 M in $C_2O_4^{2-}$.

Proceeding as in 15-12(a), we obtain $[H_3O^+] = 2.17 \times 10^{-4}$ M and pH = 3.66

(e)   Proceeding as in 15-12(b), we obtain $[H_3O^+] = 2.17 \times 10^{-4}$ and pH = 3.66

15-16. (a)   Proceeding as in 15-14(a), we obtain $[H_3O^+] = 1.287 \times 10^{-2}$ M and pH = 1.89

(b)   Recognizing that the first proton of H_2SO_4 completely dissociates we obtain

$HSO_4^- + H_2O \rightleftharpoons SO_4^{2-} + H_3O^+ \qquad K_{a2} = 1.02 \times 10^{-2}$

$$1.02 \times 10^{-2} = \frac{[H_3O^+][SO_4^{2-}]}{[HSO_4^-]} = \frac{(0.0100 + 0.0150 + x)x}{0.0150 - x}$$

Rearranging gives $x^2 + (0.0250 + 1.02 \times 10^{-2})x - (1.02 \times 10^{-2})(0.0150) = 0$

Solving the quadratic, gives $x = 3.91 \times 10^{-3}$

The total $[H_3O^+] = 0.0250 + x = 0.0289$ M and pH = 1.54

(c)   Proceeding as in 15-14(c) we obtain $[OH^-] = 0.0382$ M and pH = 12.58

(d) $CH_3COO^- + H_2O \rightleftharpoons CH_3COOH + OH^-$ $K_{b1} = \dfrac{1.00 \times 10^{-14}}{1.75 \times 10^{-5}} = 5.7 \times 10^{-10}$

CH_3COO^- is such a weak base that it makes no significant contribution to $[OH^-]$

Therefore, $[OH^-] = 0.010$ M and pH = 12.00

15-18. (a)   Proceeding as in Problem 15-16(a) with $[H_3O^+] = 1.00 \times 10^{-9}$ we obtain

$[H_2S]/[HS^-] = 0.010$

(b)   Formulating the three species as BH_2^{2+}, BH^+ and B, where B is the symbol for

$NH_2C_2H_5NH_2$.

$\dfrac{[H_3O^+][BH^+]}{[BH_2^{2+}]} = K_{a1} = 1.42 \times 10^{-7}$ and $\dfrac{[H_3O^+][B]}{[BH^+]} = K_{a2} = 1.18 \times 10^{-10}$

$[BH_2^{2+}]/[BH^+] = \dfrac{1.00 \times 10^{-9}}{1.42 \times 10^{-7}} = 0.0070$

$[B]/[BH^+] = \dfrac{1.18 \times 10^{-10}}{1.00 \times 10^{-9}} = 0.118$

$[BH_2^{2+}]$ is $< [B]$ and $[BH^+]/[B] = 1.00/0.118 = 8.5$

(c)   Proceeding as in Problem 15-16(b) we find

$[H_2AsO_4^-]/[HAsO_4^{2-}] = 9.1 \times 10^{-3}$

(d)   Proceeding as in Problem 15-16(a) we find

$[HCO_3^-]/[CO_3^{2-}] = 21$

15-20. pH = 5.75; $[H_3O^+] = $ antilog $(-5.75) = 1.778 \times 10^{-6}$

$K_{a2} = [H_3O^+][P^{2-}]/[HP^-] = 3.91 \times 10^{-6}$

$[P^{2-}]/[HP^-] = 3.91 \times 10^{-6}/(1.778 \times 10^{-6}) = 2.199$

$P^{2-} + H_2P \rightarrow 2HP^-$

amount H_2P present = 750 mL $\times$ 0.0500 M = 37.5 mmol amount HP$^-$ in the buffer $= 2 \times 37.5$ mmol $= 75.0$ mmol

amount P^{2-} needed in the buffer $= 2.199 \times 75.0$ mmol $= 164.9$ mmol

Thus, we need $37.5 + 164.9 = 202.4$ mmol of K$_2$P.

mass K$_2$P $= 202.4$ mmol $\times 0.24232$ g/mmol $= 49.0$ g

15-22. amount KHP $= 100$ mL $\times 0.150$ M $= 15.0$ mmol

(a)   amount P^{--} $= 100$ mL $\times 0.0800$ M $= 8.00$ mmol

amount KHP $= 15.0 - 8.00 = 7.00$ mmol

$c_{HP^-} = 7.00/200 = 0.0350$ M; $c_{p^{2-}} = 8.00/200 = 0.0400$ M

Proceeding as in Problem 15-12(b), we obtain pH $= 5.47$

(b) $c_{H_2P} = 8.00/200 = 0.0400$ M; $c_{HP^-} = (15.00 - 8.00)/200 = 0.0350$ M

Proceeding as in Problem 15-12(a), we obtain pH $= 2.92$

15-24. $[H_3O^+][HPO_4^{2-}]/[H_2PO_4^-] = 6.32 \times 10^{-8}$

$$\frac{[HPO_4^{2-}]}{[H_2PO_4^-]} = \frac{6.32 \times 10^{-8}}{1.00 \times 10^{-7}} = 0.632 \qquad (1)$$

Let $V_{H_3PO_4}$ and V_{NaOH} be the volume in milliliters of the two reagents. Then

$$V_{H_3PO_4} + V_{NaOH} = 1000 \text{ mL} \qquad (2)$$

From mass-balance considerations we may write that in the 1000 mL

$$\text{amount NaH}_2\text{PO}_4 + \text{amount Na}_2\text{HPO}_4 = 0.200 \times V_{H_3PO_4} \text{ mmol} \qquad (3)$$

$$\text{amount NaH}_2\text{PO}_4 + 2 \times \text{amount Na}_2\text{HPO}_4 = 0.160 \times V_{NaOH} \text{ mmol} \qquad (4)$$

Equation (1) can be rewritten

$$\frac{\text{no. mmol Na}_2\text{HPO}_4/1000}{\text{no. mmol NaH}_2\text{PO}_4/1000} = \frac{\text{no. mmol Na}_2\text{HPO}_4}{\text{no. mmol NaH}_2\text{PO}_4} = 0.632 \qquad (5)$$

Thus, we have four equations, (2), (3), (4) and (5), and four unknowns: $V_{H_3PO_4}$, V_{NaOH}, no.

mmol NaH_2PO_4 and no. mmol Na_2HPO_4. Subtracting Equation (3) from (4) yields

$$\text{no. mmol } Na_2HPO_4 = 0.160\ V_{NaOH} - 0.200\ V_{H_3PO_4} \qquad (6)$$

Substituting Equation (6) into (3) gives

$$\text{no. mmo } NaH_2PO_4 + 0.160\ V_{NaOH} - 0.200\ V_{H_3PO_4} = 0.200\ V_{H_3PO_4}$$

$$\text{no. mmo } NaH_2PO_4 = -0.160\ V_{NaOH} + 0.400\ V_{H_3PO_4} \qquad (7)$$

Substituting Equations (6) and (7) into (5) gives

$$\frac{0.160 V_{NaOH} - 0.200 V_{H_3PO_4}}{0.400 V_{H_3PO_4} - 0.160 V_{NaOH}} = 0.632$$

This equation rearranges to

$$0.2611\ V_{NaOH} = 0.4528\ V_{H_3PO_4}$$

Substituting Equation (2) gives

$$0.2611\ (1000 - V_{H_3PO_4}) = 0.4528\ V_{H_3PO_4}$$

$$V_{H_3PO_4} = 261.1/0.7139 = 366 \text{ mL} \quad \text{and} \quad V_{NaOH} = 1000 - 366 = 634 \text{ mL}$$

Thus, mix 366 mL H_3PO_4 with 634 mL NaOH

15-28. For the titration of a mixture of H_3PO_4 and $H_2PO_4^-$, the volume to the first end point

would have to be smaller than one half the total volume to the second end point because

in the titration from the first to second end points both analytes are titrated, whereas to

the first end point only the H_3PO_4 is titrated.

15-32. (a) $2H_2AsO_4^- \rightleftharpoons H_3AsO_4 + HAsO_4^{2-}$

$$K_{a1} = \frac{[H_3O^+][H_2AsO_4^-]}{[H_3AsO_4]} = 5.8 \times 10^{-3} \qquad (1)$$

$$K_{a2} = \frac{[H_3O^+][HAsO_4^{2-}]}{[H_2AsO_4^-]} = 1.1 \times 10^{-7} \qquad (2)$$

$$K_{a3} = \frac{[H_3O^+][AsO_4^{3-}]}{[HAsO_4^{2-}]} = 3.2 \times 10^{-12} \qquad (3)$$

Dividing Equation (2) by Equation (1) leads to

$$\frac{K_{a2}}{K_{a1}} = \frac{[H_3AsO_4][HAsO_4^{2-}]}{[H_2AsO_4^-]^2} = 1.9 \times 10^{-5}$$

which is the desired equilibrium constant expression.

(b) $2HAsO_4^{2-} \rightleftharpoons AsO_4^{3-} + H_2AsO_4^-$

Here we divide Equation (3) by Equation (2)

$$\frac{K_{a3}}{K_{a2}} = \frac{[AsO_4^{3-}][H_2AsO_4^-]}{[HAsO_4^{2-}]^2} = 2.9 \times 10^{-5}$$

15-34. See spreadsheet on next page.

	A	B	C	D	E	F	G	H	I	J	K
1	Pb15-34										
2	Part/Acid	pH	$[H_3O^+]$	K_{a1}	K_{a2}	K_{a3}	α_0	α_1	α_2	α_3	Alpha sum
3	(a)	2.00	1.00E-02	1.12E-03	3.91E-06		0.899	0.101	3.94E-05		1.0000000
4	phthallic	6.00	1.00E-06				1.82E-04	0.204	7.96E-01		1.0000000
5		10.00	1.00E-10				2.28E-12	2.56E-05	1.00E+00		1.0000000
6	(b)	2.00	1.00E-02	7.11E-03	6.32E-08	4.50E-13	0.584	0.416	2.63E-06	1.18E-16	1.0000000
7	phosphoric	6.00	1.00E-06				1.32E-04	0.940	5.94E-02	2.67E-08	1.0000000
8		10.00	1.00E-10				2.21E-11	1.57E-03	9.94E-01	4.47E-03	1.0000000
9	(c)	2.00	1.00E-02	7.45E-04	1.73E-05	4.02E-07	0.931	6.93E-02	1.20E-04	4.82E-09	1.0000000
10	citric	6.00	1.00E-06				5.31E-05	3.96E-02	6.85E-01	2.75E-01	1.0000000
11		10.00	1.00E-10				1.93E-16	1.44E-09	2.49E-04	1.000	1.0000000
12	(d)	2.00	1.00E-02	5.80E-03	1.10E-07	3.20E-12	0.633	0.367	4.04E-06	1.29E-15	1.0000000
13	aresenic	6.00	1.00E-06				1.55E-04	0.901	9.91E-02	3.17E-07	1.0000000
14		10.00	1.00E-10				1.52E-11	8.80E-04	9.68E-01	3.10E-02	1.0000000
15	(e)	2.00	1.00E-02	3.0E-02	1.62E-07		0.250	0.750	1.21E-05		1.0000000
16	phosphorous	6.00	1.00E-06				2.87E-05	0.861	1.39E-01		1.0000000
17		10.00	1.00E-10				2.06E-12	6.17E-04	0.999		1.0000000
18	(f)	2.00	1.00E-02	5.60E-02	5.42E-05		0.151	0.845	4.58E-03		1.0000000
19	oxalic	6.00	1.00E-06				3.23E-07	0.018	9.82E-01		1.0000000
20		10.00	1.00E-10				3.29E-15	1.85E-06	1.000		1.0000000
21											
22	**Spreadsheet Documentation**										
23	Cell C3=10^(-B3)										
24	Cell G3=$C$3^2/($C$3^2+$D$3*$C3+$D$3*$E$3)										
25	Cell H3=$C$3*$D$3/($C$3^2+$D$3*$C3+$D$3*$E$3)										
26	Cell I3=$D$3*$E$3/($C$3^2+$D$3*$C3+$D$3*$E$3)										
27	Cell K3=SUM(G3:J3)										
28	Cell G6=$C6^3/($C6^3+$D$6*$C6^2+$D$6*$E$6*$C6+$D$6*$E$6*$F$6)										
29	Cell H6=$D$6*$C6^2/($C6^3+$D$6*$C6^2+$D$6*$E$6*$C6+$D$6*$E$6*$F$6)										
30	Cell I6=$D$6*$E$6*$C6/($C6^3+$D$6*$C6^2+$D$6*$E$6*$C6+$D$6*$E$6*$F$6)										
31	Cell J6=$D$6*$E$6*$F$6/($C6^3+$D$6*$C6^2+$D$6*$E$6*$C6+$D$6*$E$6*$F$6)										

Chapter 16

16-1. Nitric acid is seldom used as a standard because it is an oxidizing agent and thus will

react with reducible species in titration mixtures.

16-3. Carbon dioxide is not strongly bonded by water molecules, and thus is readily volatilized

from aqueous solution by briefly boiling. On the other hand, HCl molecules are fully

dissociated into H_3O^+ and Cl^- when dissolved in water. Neither the H_3O^+ nor the Cl^-

species is volatile.

16-5. Let us consider the standardization of 40 mL of 0.010 M NaOH using $KH(IO_3)_2$,

$$\frac{0.010 \text{ mmol NaOH}}{\text{mL}} \times 40 \text{ mL NaOH} \times \frac{1 \text{ mmol KH(IO}_3)_2}{1 \text{ mmol NaOH}} \times \frac{390 \text{ g KH(IO}_3)_2}{1000 \text{ mmol}} = 0.16 \text{ g KH(IO}_3)_2$$

Now using benzoic acid,

$$\frac{0.010 \text{ mmol NaOH}}{\text{mL}} \times 40 \text{ mL NaOH} \times \frac{1 \text{ mmol C}_6\text{H}_5\text{COOH}}{1 \text{ mmol NaOH}} \times$$

$$\frac{122 \text{ g C}_6\text{H}_5\text{COOH}}{1000 \text{ mmol}} = 0.049 \text{ g C}_6\text{H}_5\text{COOH}$$

The primary standard $KH(IO_3)_2$ is preferable because the relative mass measurement

error would be less with a 0.16 g sample of $KH(IO_3)_2$ as opposed to 0.049 g sample of

benzoic acid. A second reason for preferring $KH(IO_3)_2$ is because it is a strong acid and

benzoic acid is not. A smaller titration error occurs when using a strong acid as a primary

standard and the choice of indicator is not critical.

16-7. If the sodium hydroxide solution is to be used for titrations with an acid-range indicator,

the carbonate in the base solution will consume two analyte hydronium ions just as would

the two hydroxides lost in the formation of Na_2CO_3.

16-9. (a)

$$\frac{0.10 \text{ mole KOH}}{L} \times 2.00 \text{ L} \times \frac{56.106 \text{ g KOH}}{\text{mole}} = 11 \text{ g KOH}$$

Dissolve 11 g KOH in water and dilute to 2.00 L total volume.

(b)

$$\frac{0.010 \text{ mole Ba(OH)}_2 \cdot 8H_2O}{L} \times 2.00 \text{ L} \times \frac{315.46 \text{ g Ba(OH)}_2 \cdot 8H_2O}{\text{mole}} = 6.3 \text{ g Ba(OH)}_2 \cdot 8H_2O$$

Dissolve 6.3 g Ba(OH)$_2 \cdot$8H$_2$O in water and dilute to 2.00 L total volume.

(c)

$$\frac{0.150 \text{ mole HCl}}{L} \times 2.00 \text{ L} \times \frac{36.461 \text{ g HCl}}{\text{mole}} \times \frac{\text{mL reagent}}{1.0579 \text{ g reagent}} \times \frac{100 \text{ g reagent}}{11.50 \text{ g HCl}} = 90 \text{ mL reagent}$$

Dilute 90 mL reagent to 2.00 L total volume.

16-11. For the first data set,

$$c_{\text{sample 1}} = \frac{0.2068 \text{ g Na}_2\text{CO}_3 \times \dfrac{1000 \text{ mmol Na}_2\text{CO}_3}{105.99 \text{ g}} \times \dfrac{2 \text{ mmol HClO}_4}{1 \text{ mmol Na}_2\text{CO}_3}}{36.31 \text{ mL HClO}_4} = 0.10747 \text{ M HClO}_4$$

The results in the accompanying table were calculated in the same way.

Sample	$c_{\text{sample i}}$, M	$c_{\text{sample i}}^2$
1	0.10747	1.15499×10^{-2}
2	0.10733	1.15196×10^{-2}
3	0.10862	1.17987×10^{-2}
4	0.10742	1.15385×10^{-2}
	$\sum c_{\text{sample i}} = 0.43084$	$\sum c_{\text{sample i}}^2 = 4.64069 \times 10^{-2}$

(a) $\overline{c}_{\text{sample i}} = \dfrac{0.43084}{4} = 0.1077 \text{ M HClO}_4$

(b)

$$s = \sqrt{\frac{(4.64069 \times 10^{-2}) - (0.43084)^2 / 4}{3}} = \sqrt{\frac{1.11420 \times 10^{-6}}{3}} = 6.1 \times 10^{-4}$$

$$CV = \frac{6.1 \times 10^{-4}}{0.1077} \times 100\% = 0.57\%$$

(c)

$$Q = \frac{0.10862 - 0.10747}{0.10862 - 0.10733} = 0.89$$

$Q_{crit} = 0.829$ at the 95% confidence level

$Q_{crit} = 0.926$ at the 99% confidence level

Thus, 0.10862 could be rejected at 95% level but must be retained at 99% level.

16-13. As in part (a) of problem 16-23,

$$c_{base} = \frac{\left(\dfrac{0.1019 \text{ mmol NaOH}}{\text{mL}} \times 500 \text{ mL}\right) - \left(0.652 \text{ g CO}_2 \times \dfrac{1000 \text{ mmol CO}_2}{44.01 \text{ g}} \times \dfrac{1 \text{ mmol NaOH}}{1 \text{ mmol CO}_2}\right)}{500 \text{ mL}}$$

$$= 0.07227 \text{ M NaOH}$$

$$\text{relative carbonate error} = \frac{0.07227 - 0.1019}{0.1019} \times 100\% = -29\%$$

16-15. (a)

$$\frac{0.1791 \text{ g BaSO}_4 \times \dfrac{1000 \text{ mmol BaSO}_4}{233.39 \text{ g}} \times \dfrac{1 \text{ mmol Ba(OH)}_2}{1 \text{ mmol BaSO}_4}}{50.00 \text{ mL Ba(OH)}_2} = 0.01535 \text{ M Ba(OH)}_2$$

(b)

$$\frac{0.4512 \text{ g KHP} \times \dfrac{1000 \text{ mmol KHP}}{204.224 \text{ g}} \times \dfrac{1 \text{ mmol Ba(OH)}_2}{2 \text{ mmol KHP}}}{26.46 \text{ mL Ba(OH)}_2} = 0.04175 \text{ M Ba(OH)}_2$$

(c)

$$\text{amnt } C_6H_5COOH = 0.3912 \text{ g } C_6H_5COOH \times \frac{1000 \text{ mmol } C_6H_5COOH}{122.123 \text{ g}} = 3.2033 \text{ mmol}$$

$$\text{amnt HCl} = \frac{0.05317 \text{ mmol HCl}}{\text{mL}} \times 4.67 \text{ mL HCl} = 0.2483 \text{ mmol}$$

$$\text{total amnt acid} = 3.2034 + 0.2483 = 3.4516 \text{ mmol}$$

$$\frac{3.4516 \text{ mmol acid} \times \dfrac{1 \text{ mmol Ba(OH)}_2}{2 \text{ mmol acid}}}{50.00 \text{ mL Ba(OH)}_2} = 0.03452 \text{ M Ba(OH)}_2$$

16-17. In Example 16-1, we found that 20.00 mL of 0.0200 M HCl requires 0.048 g TRIS, 0.021 g Na_2CO_3 and 0.08 g $Na_2B_4O_7 \cdot 10H_2O$. In each case, the absolute standard deviation in computed molar concentration of 0.0200 M HCl is

$$\text{TRIS: } s_c = \frac{0.0001}{0.048} \times 0.0200 \text{ M} = 4 \times 10^{-5} \text{ M}$$

$$Na_2CO_3 : s_c = \frac{0.0001}{0.021} \times 0.0200 \text{ M} = 1 \times 10^{-4} \text{ M}$$

$$Na_2B_4O_7 \cdot 10H_2O : s_c = \frac{0.0001}{0.076} \times 0.0200 \text{ M} = 2.5 \times 10^{-5} \text{ M} \approx 3.0 \times 10^{-5} \text{ M}$$

Proceeding as above, we calculate the relative standard deviation in the computed molar concentrations of 30.00 mL, 40.00 mL and 50.00 mL of 0.0200 M HCl and the results are shown in the table that follows.

$V_{0.0200 \text{ M HCl}}$ (mL)	Calculated masses	s_c (0.0200 M
30.00		
TRIS	0.073	3×10^{-5}
Na_2CO_3	0.032	6×10^{-5}
$Na_2B_4O_7 \cdot 10H_2O$	0.11	2×10^{-5}

40.00		
TRIS	0.097	2×10^{-5}
Na_2CO_3	0.042	5×10^{-5}
$Na_2B_4O_7 \cdot 10H_2O$	0.15	1×10^{-5}
50.00		
TRIS	0.12	2×10^{-5}
Na_2CO_3	0.053	4×10^{-5}
$Na_2B_4O_7 \cdot 10H_2O$	0.19	1×10^{-5}

16-19.

$$\text{amnt NaOH} = \frac{0.03291 \text{ mmol NaOH}}{\text{mL}} \times 24.57 \text{ mL NaOH} = 0.80860 \text{ mol NaOH}$$

$$\frac{\left(0.80860 \text{ mol NaOH} \times \dfrac{1 \text{ mmol } H_2C_4H_4O_6}{2 \text{ mmol NaOH}} \times \dfrac{150.09 \text{ g } H_2C_4H_4O_6}{1000 \text{ mmol}} \right)}{50.00 \text{ mL}} \times 100 \text{ mL}$$

$$= 0.1214 \text{ g } H_2C_4H_4O_6 \text{ per 100 mL}$$

16-21. For each part, we can write

$$\frac{\dfrac{0.1129 \text{ mmol HCl}}{\text{mL}} \times 30.79 \text{ mL HCl}}{0.7513 \text{ g sample}} = 4.6269 \frac{\text{mmol HCl}}{\text{g sample}}$$

(a)

$$4.6269 \frac{\text{mmol HCl}}{\text{g sample}} \times \frac{1 \text{ mmol } Na_2B_4O_7}{2 \text{ mmol HCl}} \times \frac{201.222 \text{ g } Na_2B_4O_7}{1000 \text{ mmol}} \times 100\% = 46.55\% \ Na_2B_4O_7$$

Proceeding in the same way

(b)

$$4.6269 \frac{\text{mmol HCl}}{\text{g sample}} \times \frac{1 \text{ mmol } Na_2B_4O_7 \cdot 10H_2O}{2 \text{ mmol HCl}} \times \frac{381.372 \text{ g } Na_2B_4O_7 \cdot 10H_2O}{1000 \text{ mmol}} \times 100\%$$

$$= 88.23\% \ Na_2B_4O_7 \cdot 10H_2O$$

(c)

$$4.6269 \; \frac{\text{mmol HCl}}{\text{g sample}} \times \frac{1 \text{ mmol B}_2\text{O}_3}{1 \text{ mmol HCl}} \times \frac{69.620 \text{ g B}_2\text{O}_3}{1000 \text{ mmol}} \times 100\% = 32.21\% \text{ B}_2\text{O}_3$$

(d)

$$4.6269 \; \frac{\text{mmol HCl}}{\text{g sample}} \times \frac{2 \text{ mmol B}}{1 \text{ mmol HCl}} \times \frac{10.811 \text{ g B}}{1000 \text{ mmol}} \times 100\% = 10.00\% \text{ B}$$

16-23.

$$\text{amnt NaOH consumed} = \left(\frac{0.0959 \text{ mmol NaOH}}{\text{mL}} \times 50.0 \text{ mL NaOH} \right) -$$

$$\left(\frac{0.05370 \text{ mmol H}_2\text{SO}_4}{\text{mL}} \times 22.71 \text{ mL H}_2\text{SO}_4 \times \frac{2 \text{ mmol NaOH}}{1 \text{ mmol H}_2\text{SO}_4} \right) = 2.356 \text{ mmol}$$

$$\frac{2.356 \text{ mmol NaOH} \times \dfrac{1 \text{ mmol HCHO}}{1 \text{ mmol NaOH}} \times \dfrac{30.026 \text{ g HCHO}}{1000 \text{ mmol}}}{0.2985 \text{ g sample}} \times 100\% = 23.7\% \text{ HCHO}$$

16-25. Tetraethylthiuram disulfide, TS$_4$

$$1 \text{ mmol TS}_4 \equiv 4 \text{ mmol SO}_2 \equiv 4 \text{ mmol H}_2\text{SO}_4 \equiv 8 \text{ mmol NaOH}$$

$$\frac{\left(\dfrac{0.04216 \text{ mmol NaOH}}{\text{mL}} \times 19.25 \text{ mL NaOH} \times \dfrac{1 \text{ mmol TS}_4}{8 \text{ mmol NaOH}} \times \dfrac{296.54 \text{ g TS}_4}{1000 \text{ mmol}} \right)}{0.4169 \text{ g sample}} \times 100\%$$

$$= 7.216\% \text{ TS}_4$$

16-27.

$$\text{amnt HCl} = \text{mmol NaOH} - 2 \times \text{mmol CO}_3^{2-}$$

$$\text{amnt CO}_3^{2-} = \frac{\left(\dfrac{0.1140 \text{ mmol HCl}}{\text{mL}} \times 50.00 \text{ mL HCl} \right) - \left(\dfrac{0.09802 \text{ mmol NaOH}}{\text{mL}} \times 24.21 \text{ mL NaOH} \right)}{2}$$

$$= 1.6635 \text{ mmol CO}_3^{2-}$$

$$\text{molar mass carbonate salt} = \frac{0.1401 \text{ g salt}}{1.6635 \text{ mmol CO}_3^{2-}} \times \frac{1000 \text{ mmol}}{\text{mole}} = 84.22 \; \frac{\text{g salt}}{\text{mole CO}_3^{2-}}$$

molar mass of carbonate salt cation $= \left(84.22 \dfrac{\text{g salt}}{\text{mole CO}_3{}^{2-}} \times \dfrac{1 \text{ mole CO}_3{}^{2-}}{1 \text{ mole salt}} \right) - 60.01 \dfrac{\text{g CO}_3{}^{2-}}{\text{mole}}$

$= 24.21 \dfrac{\text{g cation}}{\text{mole}}$

$MgCO_3$ with a molar mass of 84.31 g/mole appears to be a likely candidate

16-29.

amnt $Ba(OH)_2 = $ mmol $CO_2 + \dfrac{\text{mmol HCl}}{2}$

amnt $CO_2 = \left(\dfrac{0.0116 \text{ mmol Ba(OH)}_2}{\text{mL}} \times 50.0 \text{ mL Ba(OH)}_2 \right) - \left(\dfrac{\dfrac{0.0108 \text{ mmol HCl}}{\text{mL}} \times 23.6 \text{ mL HCl}}{2} \right)$

$= 4.526 \times 10^{-1}$ mmol

$\dfrac{0.4526 \text{ mmol CO}_2 \times \dfrac{44.01 \text{ g CO}_2}{1000 \text{ mmol}}}{3.00 \text{ L}} \times \dfrac{1 \text{ L CO}_2}{1.98 \text{ g CO}_2} \times 10^6 \text{ppm} = 3.35 \times 10^3 \text{ppm CO}_2$

16-31. $(NH_4)_3PO_4 \cdot 12MoO_3(s) + 26OH^- \rightarrow HPO_4{}^{2-} + 12MoO_4{}^{2-} + 14H_2O + 3NH_3(g)$

amnt NaOH consumed $= \left(\dfrac{0.2000 \text{ mmol NaOH}}{\text{mL}} \times 50.00 \text{ mL NaOH} \right) -$

$\left(\dfrac{0.1741 \text{ mmol HCl}}{\text{mL}} \times 14.17 \text{ mL HCl} \right) = 7.533$ mmol

amnt P $= 7.533 \text{ mmol NaOH} \times \dfrac{1 \text{ mmol (NH}_4)_3\text{PO}_4 \cdot 12\text{MoO}_3}{26 \text{ mmol NaOH}} \times$

$\dfrac{1 \text{ mmol P}}{1 \text{ mmol (NH}_4)_3\text{PO}_4 \cdot 12\text{MoO}_3} = 2.897 \times 10^{-1}$ mmol

$\dfrac{2.897 \times 10^{-1} \text{ mmol P} \times \dfrac{30.974 \text{ g P}}{1000 \text{ mmol}}}{0.1417 \text{ g sample}} \times 100\% = 6.333\% \text{ P}$

121

16-33.

Neohetramine, $C_{16}H_{21}ON_4 = RN_4$

1 mmol $RN_4 \equiv 3$ mmol $NH_3 \equiv 4$ mmol HCl

$$\frac{\dfrac{0.01477 \text{ mmol HCl}}{\text{mL}} \times 26.13 \text{ mL HCl} \times \dfrac{1 \text{ mmol } RN_4}{4 \text{ mmol HCl}} \times \dfrac{285.37 \text{ g } RN_4}{1000 \text{ mmol}}}{0.1247 \text{ g sample}} \times 100\% = 22.08\% \ RN_4$$

16-35.

$$\%N = \frac{\left(\dfrac{0.1249 \text{ mmol HCl}}{\text{mL}} \times 20.59 \text{ mL HCl} \right) \times \dfrac{1 \text{ mmol N}}{\text{mmol HCl}} \times \dfrac{14.007 \text{ g N}}{1000 \text{ mmol}}}{0.917 \text{ g sample}} \times 100\% = 3.93\% \ N$$

16-37.

	A	B	C	D	E	F
1	**Problem 16-37**					
2	Mass sample, g	0.5843				
3	Vol. HCl, mL	50.00				
4	Conc. HCl, M	0.1062				
5	Vol. NaOH, mL	11.89				
6	Conc. NaOH, M	0.0925				
7						
8	amnt HCl/g, mmol/g	7.20550231	amnt HCl/g = (mmol HCl - mmol NaOH)/sample mass			
9		**Molar masses**	**Percentages**			
10	(a) N	14.007	10.09			
11	(b) urea	60.06	21.64			
12	(c) $(NH_4)_2SO_4$	132.141	47.61			
13	(d) $(NH_4)_3PO_4$	149.09	35.81			
14	**Spreadsheet Documentation**					
15	Cell B8=(B3*B4-B5*B6)/B2		The percentages are calculated in Cells C10:C13			
16	Cell C10=$B$8*1*B10/1000*100		from the no. of mmol of HCl/g times the no. of mmol			
17	Cell C11=$B$8*1/2*B11/1000*100		of the compound/mmol HCl times the molar mass			
18	Cell C12=$B$8*1/2*B12/1000*100		of the compound divided by 1000 (mmolar mass).			
19	Cell C13=$B$8*1/3*B13/1000*100					
20						

16-39. In the first titration,

$$\text{amnt HCl consumed} = \left(\frac{0.08421 \text{ mmol HCl}}{\text{mL}} \times 30.00 \text{ mL} \right) -$$

$$\left(\frac{0.08802 \text{ mmol NaOH}}{\text{mL}} \times 10.17 \text{ mL} \right) = 1.63114 \text{ mmol}$$

and

$$1.63114 \text{ mmol HCl} = \text{mmol } NH_4NO_3 + \left(2 \times \text{mmol } (NH_4)_2SO_4 \right)$$

The amounts of the two species in the original sample are

$$\text{mmol } NH_4NO_3 + \left(2 \times \text{mmol } (NH_4)_2SO_4 \right) = 1.63114 \text{ mmol} \times \frac{200 \text{ mL}}{50 \text{ mL}} = 6.5246 \text{ mmol} \ (1)$$

In the second titration,

$$\text{amnt HCl consumed} = \left(\frac{0.08421 \text{ mmol HCl}}{\text{mL}} \times 30.00 \text{ mL} \right) -$$

$$\left(\frac{0.08802 \text{ mmol NaOH}}{\text{mL}} \times 14.16 \text{ mL} \right) = 1.27994 \text{ mmol HCl}$$

and

$$1.27994 \text{ mmol HCl} = (2 \times \text{mmol } NH_4NO_3) + (2 \times \text{mmol } (NH_4)_2SO_4)$$

The amounts of the two species in the original sample are

$$(2 \times \text{mmol } NH_4NO_3) + (2 \times \text{mmol } (NH_4)_2SO_4) = 1.27994 \text{ mmol} \times$$

$$\frac{200 \text{ mL}}{25 \text{ mL}} = 10.2395 \text{ mmol} \tag{2}$$

Subtracting equation (1) from equation (2) gives

$$\text{amnt } NH_4NO_3 = 10.2395 \text{ mmol} - 6.52455 \text{ mmol} = 3.7149 \text{ mmol}$$

$$\text{amnt } (NH_4)_2SO_4 = \frac{10.2395 \text{ mmol} - (2 \times 3.7149 \text{ mmol})}{2} = 1.4048 \text{ mmol}$$

$$\text{percentage } NH_4NO_3 = \frac{3.7149 \text{ mmol } NH_4NO_3 \times \dfrac{80.04 \text{ g } NH_4NO_3}{1000 \text{ mmol}}}{1.219 \text{ g sample}} \times 100\% = 24.39\%$$

$$\text{percentage } (NH_4)_2SO_4 = \frac{1.4048 \text{ mmol } (NH_4)_2SO_4 \times \dfrac{132.14 \text{ g } (NH_4)_2SO_4}{1000 \text{ mmol}}}{1.219 \text{ g sample}} \times 100\%$$

$$= 15.23\%$$

16-41. For the first aliquot,

amnt HCl = mmol NaOH + mmol $NaHCO_3$ + (2 × mmol Na_2CO_3)

$$mmol\ NaHCO_3 + (2 \times mmol\ Na_2CO_3) = \left(\frac{0.01255\ mmol\ HCl}{mL} \times 50.00\ mL\ HCl \right) -$$

$$\left(\frac{0.01063\ mmol\ NaOH}{mL} \times 2.34\ mL\ NaOH \right) = 0.6026\ mmol$$

For the second aliquot,

amnt $NaHCO_3$ = mmol NaOH − mmol HCl

$$= \left(\frac{0.01063\ mmol\ NaOH}{mL} \times 25.00\ mL\ NaOH \right) - \left(\frac{0.01255\ mmol\ HCl}{mL} \times 7.63\ mL\ HCl \right)$$

$$= 0.1700\ mmol$$

$$percentage\ NaHCO_3 = \frac{0.1700\ mmol\ NaHCO_3 \times \dfrac{84.01\ g\ NaHCO_3}{1000\ mmol}}{\left(0.5000\ g \times \dfrac{25.00\ g}{250.0\ g} \right)} \times 100\% = 28.56\%$$

$$percentage\ Na_2CO_3 = \frac{0.2163\ mmol\ Na_2CO_3 \times \dfrac{105.99\ g\ Na_2CO_3}{1000\ mmol}}{\left(0.5000\ g \times \dfrac{25.00\ mL}{250.0\ mL} \right)} \times 100\% = 45.85\%$$

$$100\% - (28.56\% + 45.85\%) = 25.59\%\ H_2O$$

125

16-43.

	A	B	C	D	E
1	**Problem 16-43**				
2	Conc. NaOH, M	0.07731			
3	(a) and (b)				
4	Conc. HCl, M	0.03000			
5	Conc. H_3PO_4, M	0.01000			
6	**(a)**		**Vol. solution, mL**	**Amnt. acid, mmol**	**Vol. NaOH, mL**
7	React with 1 proton		25.00	1.0000	12.93
8	to bromocresol green				
9	end point				
10	**(b)**		**Vol. solution, mL**	**Amnt. acid, mmol**	**Vol. NaOH, mL**
11	React with 2 protons		25.00	1.2500	16.17
12	to thymolphtalein				
13	end point				
14	**(c)**		**Vol. solution, mL**	**Amnt. acid, mmol**	**Vol. NaOH, mL**
15	Conc. NaH_2PO_4, M	0.06407	10.00	0.6407	8.29
16			20.00	1.2814	16.57
17			30.00	1.9221	24.86
18			40.00	2.5628	33.15
19	**(d) Mixture**		**Vol. solution, mL**	**Amnt. acid, mmol**	**Vol. NaOH, mL**
20	Conc. H_3PO_4, M	0.02000	20.00	1.4000	18.11
21	Conc. NaH_2PO_4, M	0.03000	25.00	1.7500	22.64
22	React with 2 protons		30.00	2.1000	27.16
23	from H_3PO_4 and 1				
24	proton from NaH_2PO_4				
25					
26	**Spreadsheet Documentation**				
27	Cell D7=C7*$B$4+C7*$B$5				
28	Cell E7=D7/$B$2				
29	Cell D11=C11*$B$4+2*C11*$B$5				
30	Cell E11=D11/$B$2				
31	Cell D15=$B$15*C15				
32	Cell E15=D15/$B$2				
33	Cell D20=2*$B$20*C20+$B$21*C20				
34	Cell E20=D20/$B$2				

16-45.

	A	B	C	D	E	F
1	Problem 16-45			Vol. to phenol, mL	Vol. to BCG, mL	
2	Conc. HCl, M	0.1202	(a)	22.42	22.44	
3	Vol. solution, mL	25.00	(b)	15.67	42.13	
4	$\mathcal{M}$ NaOH	39.997	(c)	29.64	36.42	
5	$\mathcal{M}$ Na$_2$CO$_3$	105.989	(d)	16.12	32.23	
6	$\mathcal{M}$ NaHCO$_3$	84.007	(e)	0.00	33.33	
7	Table 16-2 gives the volume relationships in titrations of these mixtures					
9	(a) Since essentially the same volue is used for both end points, there is only NaOH					
10	present. We use the average volume to calculate the concentration of NaOH in mg/mL.					
11		Amnt NaOH, mmol	conc. NaOH, mg/mL			
12		2.6961	4.313			
13	(b) Since $V_{phth} < \frac{1}{2}V_{bog}$, only carbonate and bicarbonate are present.					
14		Amnt carbonate, mmol	Amnt total, mmol	Amnt bicarbonate, mmol	Conc. bicarb., mg/mL	Conc. carb., mg/mL
15		1.8835	5.0640	1.2970	7.985	4.358
16	(c) Now $V_{phth} > \frac{1}{2}V_{bog}$, so we have a mixture of NaOH and Na$_2$CO$_3$					
17		Amnt carb. + NaOH, mmol	Amnt carb., mmol	Amnt NaOH, mmol	Conc. Na$_2$CO$_3$, mg/mL	Conc. NaOH, mg/mL
18		3.5627	0.8150	2.7478	3.455	4.396
19	(d) Since $V_{phth} = \frac{1}{2}V_{bog}$, we have only Na2CO3 present.					
20		Ammt carbonate, mmol	Conc. Na$_2$CO$_3$, mg/mL			
21		1.9376	8.215			
22	(e) Since $V_{phth} = 0$, we have only NaHCO$_3$ present which gains one proton.					
23		Amnt NaHCO$_3$, mmol	Conc. NaHCO$_3$, mg/mL			
24		4.0063	13.462			
25	Spreadsheet Documentation					
26	Cell B12=((D2+E2)/2)*$B$2		Cdll C18=($E$4-$D$4)*$B$2			
27	Cell C12=B12*1*$B$4/$B$3		Cell D18=B18-C18			
28	Cell B15=D3*$B$2		Cell E18=C18*$B$5/$B$3			
29	Cell C15=E3*$B$2		Cell F18=D18*$B$4/$B$3			
30	Cell D15=C15-2*B15		Cell B21=$D$5*$B$2			
31	Cell E15=B15*$B$5/$B$3		Cell C21=B21*1*$B$5/$B$3			
32	Cell F15=D15*$B$6/$B$3		Cell B24=E6*$B$2			
33	Cell B18=D4*$B$2		Cell C24=B24*1*$B$6/$B$3			

16-47. (a) With bromocresol green, only one of the two protons in the oxalic acid will react.

Therefore, the equivalent mass is the molar mass, or 126.066 g.

(b) When phenolphthalein is the indicator, two of the protons are consumed. Therefore,

the equivalent mass of oxalic acid is one-half the molar mass, or 63.03 g.

Chapter 17

17-1. **(a)** A *ligand* is a species that contains one or more electron pair donor groups that tend to

form bonds with metal ions.

(c) A *tetradentate chelating agent* is a molecule that contains four pairs of donor electron

located in such positions that they all can bond to a metal ion, thus forming two rings.

(e) *Argentometric titrations* are titrations based on the formation of precipitates with

standard solutions of silver nitrate. An example is the titration of a halide ion with silver

nitrate to form the isoluble silver halide.

(g) In an *EDTA displacement titration*, an unmeasured excess of a solution containing

the magnesium or zinc complex of EDTA is introduced into the solution of an analyte

that forms a more stable complex that that of magnesium or zinc. The liberated

magnesium or zinc ions are then titrated with a standard solution of EDTA.

Displacement titrations are used for the determination of cations for which no good

indicator exists.

17-2. Three general methods for performing EDTA titrations are (1) direct titration, (2) back

titration, and (3) displacement titration. Method (1) is simple, rapid, but requires one

standard reagent. Method (2) is advantageous for those metals that react so slowly with

EDTA as to make direct titration inconvenient. In addition, this procedure is useful for

cations for which satisfactory indicators are not available. Finally, it is useful for

analyzing samples that contain anions that form sparingly soluble precipitates with the

analyte under analytical conditions. Method (3) is particularly useful in situations where

no satisfactory indicators are available for direct titration.

17-3. (a)

$$Ag^+ + S_2O_3^{2-} \rightleftharpoons Ag(S_2O_3)^- \qquad\qquad K_1 = \frac{[Ag(S_2O_3)^-]}{[Ag^+][S_2O_3^{2-}]}$$

$$Ag(S_2O_3)^- + S_2O_3^{2-} \rightleftharpoons Ag(S_2O_3)_2^{3-} \qquad\qquad K_2 = \frac{[Ag(S_2O_3)_2^{3-}]}{[Ag(S_2O_3)^-][S_2O_3^{2-}]}$$

17-4. The overall formation constant β_n is equal to the product of the individual stepwise constants. Thus, the overall constant for formation of $Ag(S_2O_3)_2^{3-}$ in Problem 17-3 (a) is

$$\beta_2 = K_1K_2 = \frac{[Ag(S_2O_3)_2^{3-}]}{[Ag^+][S_2O_3^{2-}]^2}$$

17-5. The Fajans determination of chloride involves a direct titration, while a Volhard titration requires two standard solutions and a flitration step to remove AgCl before back titration of the excess SCN$^-$.

17-6. The ions that are preferentially absorbed on the surface of an ionic solid are generally lattice ions. Thus, in the beginning stages of a precipitation titration, one of the lattice ions is in excess and its charge determines the sign of the charge of the particles. After the equivalence point, the ion of the opposite charge is present in excess and determines the sign of the charge on the particle. Thus, in the equivalence-point region, the charge shift from positive to negative, or the reverse.

17-7. (a) Acetate (OAc$^-$)

$$HOAc \rightleftharpoons OAc^- + H^+ \qquad\qquad K_a = \frac{[OAc^-][H^+]}{[HOAc]}$$

$$c_T = [\text{HOAc}] + [\text{OAc}^-]$$

$$= \frac{[\text{OAc}^-][\text{H}^+]}{K_a} + [\text{OAc}^-] = [\text{OAc}^-]\left\{\frac{[\text{H}^+]}{K_a} + 1\right\} = [\text{OAc}^-]\left\{\frac{[\text{H}^+] + K_a}{K_a}\right\}$$

$$\alpha_1 = \frac{[\text{OAc}^-]}{c_T} = \frac{K_a}{[\text{H}^+] + K_a}$$

(b) Tartrate (T^{2-})

$$\text{H}_2\text{T} \rightleftharpoons \text{HT}^- + \text{H}^+ \qquad\qquad K_{a1} = \frac{[\text{HT}^-][\text{H}^+]}{[\text{H}_2\text{T}]}$$

$$\text{HT}^- \rightleftharpoons \text{T}^{2-} + \text{H}^+ \qquad\qquad K_{a2} = \frac{[\text{T}^{2-}][\text{H}^+]}{[\text{HT}^-]}$$

$$c_T = [\text{H}_2\text{T}] + [\text{HT}^-] + [\text{T}^{2-}]$$

$$= \frac{[\text{HT}^-][\text{H}^+]}{K_{a1}} + \frac{[\text{T}^{2-}][\text{H}^+]}{K_{a2}} + [\text{T}^{2-}] = \frac{[\text{T}^{2-}][\text{H}^+]^2}{K_{a1}K_{a2}} + \frac{[\text{T}^{2-}][\text{H}^+]}{K_{a2}} + [\text{T}^{2-}]$$

$$= [\text{T}^{2-}]\left\{\frac{[\text{H}^+]^2}{K_{a1}K_{a2}} + \frac{[\text{H}^+]}{K_{a2}} + 1\right\} = [\text{T}^{2-}]\left\{\frac{[\text{H}^+]^2 + K_{a1}[\text{H}^+] + K_{a1}K_{a2}}{K_{a1}K_{a2}}\right\}$$

$$\alpha_2 = \frac{[\text{T}^{2-}]}{c_T} = \frac{K_{a1}K_{a2}}{[\text{H}^+]^2 + K_{a1}[\text{H}^+] + K_{a1}K_{a2}}$$

(c) Phosphate

$$\text{H}_3\text{PO}_4 \rightleftharpoons \text{H}_2\text{PO}_4^- + \text{H}^+ \qquad\qquad K_{a1} = \frac{[\text{H}_2\text{PO}_4^-][\text{H}^+]}{[\text{H}_3\text{PO}_4]}$$

$$\text{H}_2\text{PO}_4^- \rightleftharpoons \text{HPO}_4^{2-} + \text{H}^+ \qquad\qquad K_{a2} = \frac{[\text{HPO}_4^{2-}][\text{H}^+]}{[\text{H}_2\text{PO}_4^-]}$$

$$\text{HPO}_4^{2-} \rightleftharpoons \text{PO}_4^{3-} + \text{H}^+ \qquad\qquad K_{a3} = \frac{[\text{PO}_4^{3-}][\text{H}^+]}{[\text{HPO}_4^{2-}]}$$

$$c_T = [H_3PO_4] + [H_2PO_4^-] + [HPO_4^{2-}] + [PO_4^{3-}]$$

Proceeding as in the preceeding problem, we obtain

$$c_T = [PO_4^{3-}] \left\{ \frac{[H^+]^3 + K_{a1}[H^+]^2 + K_{a1}K_{a2}[H^+] + K_{a1}K_{a2}K_{a3}}{K_{a1}K_{a2}K_{a3}} \right\}$$

$$\alpha_3 = \frac{[PO_4^{3-}]}{c_T} = \frac{K_{a1}K_{a2}K_{a3}}{[H^+]^3 + K_{a1}[H^+]^2 + K_{a1}K_{a2}[H^+] + K_{a1}K_{a2}K_{a3}}$$

17-8.

$$Fe^{3+} + 3Ox^{2-} \rightleftharpoons Fe(Ox)_3^{3-} \qquad \beta_3 = \frac{[Fe(Ox)_3^{3-}]}{[Fe^{3+}][Ox^{2-}]^3}$$

$$\alpha_2 = \frac{[Ox^{2-}]}{c_T} \qquad \text{so } [Ox^{2-}] = \alpha_2 c_T$$

$$\beta_3 = \frac{[Fe(Ox)_3^{3-}]}{[Fe^{3+}][Ox^{2-}]^3} = \frac{[Fe(Ox)_3^{3-}]}{[Fe^{3+}](\alpha_2 c_T)^3}$$

$$\beta_3' = (\alpha_2)^3 \beta_3 = \frac{[Fe(Ox)_3^{3-}]}{[Fe^{3+}](c_T)^3}$$

17-9.

$$\beta_n = \frac{[ML_n]}{[M][L]^n}$$

Taking the logarithm of both sides of the above equation yields

$$\log \beta_n = \log[ML_n] - \log[M] - n\log[L]$$

Now write the right hand side of the equation as a p function (i.e. pM = $-\log$[M]).

$$\log \beta_n = pM + npL - pML_n$$

17-10.

$$\frac{3.426 \ \cancel{g \ reagent} \times \dfrac{99.7 \ \cancel{g \ Na_2H_2Y \cdot 2H_2O}}{100 \ \cancel{g \ reagent}} \times \dfrac{1 \ mole \ EDTA}{372.24 \ \cancel{g \ Na_2H_2Y \cdot 2H_2O}}}{1.000 \ L} = 0.00918 \ M \ EDTA$$

17-11. First calculate the $CoSO_4$ concentration

$$\frac{1.569 \ \cancel{mg}}{mL} \times \frac{1 \ mmol \ CoSO_4}{155.0 \ \cancel{mg}} = 0.010123 \ M$$

In each part 25.00 mL of this solution is taken, so

$$amount \ CoSO_4 \ = \ 25.00 \ mL \times \frac{0.010123 \ mmol}{mL} = 0.25306 \ mmol$$

(a)

$$Vol. \ EDTA = 0.25306 \ \cancel{mmol \ CoSO_4} \times \frac{1 \ \cancel{mmol \ EDTA}}{\cancel{mmol \ CoSO_4}} \times \frac{1 \ mL}{0.007840 \ \cancel{mmol \ EDTA}} = 32.28 \ mL$$

(b)

$$amnt \ excess \ EDTA = \left(\frac{0.007840 \ mmol}{mL} \times 50.00 \ mL \right)$$
$$- \left(0.25306 \ mmol \ CoSO_4 \times \frac{1 \ mmol}{mmol \ CoSO_4} \right) = 0.1389 \ mmol$$

$$Vol. \ Zn^{2+} = 0.1389 \ \cancel{mmol \ EDTA} \times \frac{1 \ \cancel{mmol \ Zn^{2+}}}{\cancel{mmol \ EDTA}} \times \frac{1 \ mL}{0.009275 \ \cancel{mmol \ Zn^{2+}}} = 14.98 \ mL$$

(c)

$$Vol. \ EDTA = 0.25306 \ \cancel{mmol \ CoSO_4} \times \frac{1 \ \cancel{mmol \ Zn^{2+}}}{\cancel{mmol \ CoSO_4}} \times \frac{1 \ \cancel{mmol \ EDTA}}{\cancel{mmol \ Zn^{2+}}} \times \frac{1 \ mL}{0.007840 \ \cancel{mmol \ EDTA}}$$
$$= 32.28 \ mL$$

17-12. (a)

$$\text{Vol. EDTA} = \frac{0.0598 \ \text{mmol Mg(NO}_3)_2}{\text{mL}} \times 29.13 \ \text{mL} \times \frac{1 \ \text{mmol EDTA}}{\text{mmol Mg(NO}_3)_2} \times \frac{\text{mL}}{0.0500 \ \text{mmol EDTA}}$$

$$= 34.84 \ \text{mL}$$

(c)

$$\text{Amnt. CaHPO}_4 \cdot 2H_2O = 0.4861 \ \text{g} \times \frac{81.4 \ \text{g CaHPO}_4 \cdot 2H_2O}{100 \ \text{g}} \times \frac{1000 \ \text{mmol}}{172.09 \ \text{g CaHPO}_4 \cdot 2H_2O}$$

$$= 2.2993 \ \text{mmol}$$

$$\text{Vol. EDTA} = 2.2993 \ \text{mmol CaHPO}_4 \cdot 2H_2O \times \frac{1 \ \text{mmol EDTA}}{\text{mmol CaHPO}_4 \cdot 2H_2O} \times \frac{1 \ \text{mL}}{0.0500 \ \text{mmol EDTA}}$$

$$= 45.99 \ \text{mL}$$

(e)

$$\text{Vol. EDTA} = 0.1612 \ \text{g} \times \frac{92.5 \ \text{g}}{100 \ \text{g}} \times \frac{1000 \ \text{mmol dolo}}{184.4 \ \text{g}} \times \frac{2 \ \text{mmol EDTA}}{\text{mmol dolo}} \times \frac{1 \ \text{mL}}{0.0500 \ \text{mmol EDTA}}$$

$$= 32.34 \ \text{mL}$$

17-13.

$$\text{Wt. Zn} = \frac{0.01639 \ \text{mmol EDTA}}{\text{mL}} \times 22.57 \ \text{mL} \times \frac{1 \ \text{mmol Zn}^{2+}}{\text{mmol EDTA}} \times \frac{65.39 \ \text{g}}{1000 \ \text{mmol Zn}^{2+}} = 0.024189 \ \text{g}$$

$$\text{Percentage Zn} = \frac{0.024189 \ \text{g Zn}}{0.7457 \ \text{g sample}} \times 100\% = 3.244\%$$

17-14. Conc. $AgNO_3 = \dfrac{14.77 \text{ g}}{L} \times \dfrac{1 \text{ mol } AgNO_3}{169.873 \text{ g}} = 0.08695$ M

(a)

Vol. $AgNO_3 = 0.2631 \text{ g} \times \dfrac{\text{mmol NaCl}}{0.05833 \text{ g}} \times \dfrac{1 \text{ mmol } AgNO_3}{\text{mmol NaCl}} \times \dfrac{1 \text{ mL } AgNO_3}{0.08695 \text{ mmol } AgNO_3} = 51.78$ mL

(c)

$V_{AgNO_3} = 64.13 \text{ mg} \times \dfrac{\text{mmol } Na_3AsO_4}{207.89 \text{ mg}} \times \dfrac{3 \text{ mmol } AgNO_3}{\text{mmol } Na_3AsO_4} \times \dfrac{1 \text{ mL}}{0.08695 \text{ mmol } AgNO_3} = 10.64$ mL

(e)

$V_{AgNO_3} = 25.00 \text{ mL} \times \dfrac{0.05361 \text{ mmol } Na_3PO_4}{\text{mL}} \times \dfrac{3 \text{ mmol } AgNO_3}{\text{mmol } Na_3PO_4} \times \dfrac{1 \text{ mL}}{0.08695 \text{ mmol } AgNO_3} = 46.24$ mL

17-15. (a) An excess is assured if the calculation is based on a pure sample.

Vol. $AgNO_3 = 0.2513 \text{ g} \times \dfrac{1 \text{ mmol NaCl}}{0.05844 \text{ g}} \times \dfrac{1 \text{ mmol } AgNO_3}{\text{mmol NaCl}} \times \dfrac{1 \text{ mL}}{0.09621 \text{ mmol } AgNO_3} = 44.70$ mL

(c)

Vol. $AgNO_3 = 25.00 \text{ mL} \times \dfrac{0.01907 \text{ mmol } AlCl_3}{\text{mL}} \times \dfrac{3 \text{ mmol } AgNO_3}{\text{mmol } AlCl_3} \times \dfrac{1 \text{ mL}}{0.09621 \text{ mmol } AgNO_3} = 14.87$ mL

17-16.

Percent $Tl_2SO_4 = \dfrac{\left(\dfrac{0.03610 \text{ mmol EDTA}}{\text{mL}} \times 12.77 \text{ mL} \times \dfrac{1 \text{ mmol } Tl_2SO_4}{2 \text{ mmol EDTA}} \times \dfrac{504.8 \text{ g}}{\text{mmol } Tl_2SO_4} \right)}{9.57 \text{ g sample}} \times 100\%$

$= 1.216\%$

17-17.

$$\text{Amnt Fe}^{3+} = \frac{0.01500 \; \frac{\text{mmol EDTA}}{\text{mL}}}{} \times 10.98 \; \text{mL} \times \frac{1 \; \text{mmol Fe}^{3+}}{\text{mmol EDTA}} = 0.1647 \; \text{mmol}$$

$$\text{Amnt Fe}^{2+} = \frac{0.01500 \; \frac{\text{mmol EDTA}}{\text{mL}}}{} \times (23.70 - 10.98) \; \text{mL} \times \frac{1 \; \text{mmol Fe}^{2+}}{\text{mmol EDTA}} = 0.1908 \; \text{mmol}$$

$$\text{Conc. Fe}^{3+} = \frac{\left(0.1647 \; \text{mmol Fe}^{3+} \times \frac{55.847 \; \text{mg}}{\text{mmol Fe}^{3+}} \right)}{50.00 \; \text{mL} \times \frac{\text{L}}{1000 \; \text{mL}}} = 183.96 \; \text{ppm} \approx 184.0 \; \text{ppm}$$

$$\text{Conc. Fe}^{2+} = \frac{\left(0.1908 \; \text{mmol Fe}^{2+} \times \frac{55.847 \; \text{mg}}{\text{mmol Fe}^{2+}} \right)}{50.00 \; \text{mL} \times \frac{\text{L}}{1000 \; \text{mL}}} = 213.1 \; \text{ppm}$$

17-18.

Amount $Cd^{2+} + Pb^{2+} =$

$$\frac{0.06950 \text{ mmol EDTA}}{\text{mL}} \times 28.89 \text{ mL EDTA} \times \frac{1 \text{ mmol } (Cd^{2+} + Pb^{2+})}{\text{mmol EDTA}} = 2.00786 \text{ mmol}$$

$$\text{Amnt } Pb^{2+} = \frac{0.06950 \text{ mmol EDTA}}{\text{mL}} \times 11.56 \text{ mL EDTA} \times \frac{1 \text{ mmol } Pb^{2+}}{\text{mmol EDTA}} = 0.80342 \text{ mmol}$$

$$\text{Amnt } Cd^{2+} = 2.00786 \text{ mmol} - 0.80342 \text{ mmol} = 1.20444 \text{ mmol}$$

$$\frac{\left(0.80342 \text{ mmol } Pb^{2+} \times \dfrac{207.2 \text{ g } Pb^{2+}}{1000 \text{ mmol}} \right)}{1.509 \text{ g sample} \times \dfrac{50.00 \text{ mL}}{250.0 \text{ mL}}} \times 100\% = 55.16\% \ Pb^{2+}$$

$$\frac{\left(1.204 \text{ mmol } Cd^{2+} \times \dfrac{112.41 \text{ g } Cd^{2+}}{1000 \text{ mmol}} \right)}{1.509 \text{ g sample} \times \dfrac{50.00 \text{ mL}}{250.0 \text{ mL}}} \times 100\% = 44.86\% \ Cd^{2+}$$

17-19.

$$\frac{\left(\dfrac{0.01133 \text{ mmol EDTA}}{\text{mL}} \times 38.37 \text{ mL EDTA} \times \dfrac{1 \text{ mmol ZnO}}{\text{mmol EDTA}} \times \dfrac{81.379 \text{ g ZnO}}{1000 \text{ mmol}} \right)}{1.056 \text{ g sample} \times \dfrac{10.00 \text{ mL}}{250.0 \text{ mL}}} \times 100\%$$

$= 83.75\%$ ZnO

$$\frac{\left(\dfrac{0.002647 \text{ mmol } ZnY^{2-}}{\text{mL}} \times 2.30 \text{ mL } ZnY^{2-} \times \dfrac{1 \text{ mmol } Fe_2O_3}{2 \text{ mmol } ZnY^{2-}} \times \dfrac{159.69 \text{ g } Fe_2O_3}{1000 \text{ mmol}} \right)}{1.056 \text{ g sample} \times \dfrac{50.00 \text{ mL}}{250.0 \text{ mL}}} \times 100\%$$

$= 0.230\%$ Fe_2O_3

17-20.

$$1 \text{ mmol EDTA} \equiv 1 \text{ mmol Ni}^{2+} \equiv 2 \text{ mmol NaBr} \equiv 2 \text{ mmol NaBrO}_3$$

For the 10.00 mL aliquot,

$$\frac{\text{Amnt NaBr} + \text{amnt NaBrO}_3}{\text{mL sample solution}} =$$

$$\frac{\left(\dfrac{0.02089 \text{ mmol EDTA}}{\text{mL}} \times 21.94 \text{ mL EDTA} \times \dfrac{2\left(\text{mmol NaBr} + \text{mmol NaBrO}_3\right)}{\text{mmol EDTA}} \right)}{10.00 \text{ mL}} = 0.09166 \text{ M}$$

For the 25.00 mL aliquot,

$$\frac{\text{Amnt NaBr}}{\text{mL sample solution}} =$$

$$\frac{\left(\dfrac{0.02089 \text{ mmol EDTA}}{\text{mL}} \times 26.73 \text{ mL EDTA} \times \dfrac{2 \text{ mmol NaBr}}{\text{mmol EDTA}} \right)}{25.00 \text{ mL}} = 0.04467 \text{ M NaBr}$$

$$\frac{\text{Amnt NaBrO}_3}{\text{mL sample solution}} = 0.09166 - 0.04467 = 0.04699 \text{ M NaBrO}_3$$

$$\frac{\left(\dfrac{0.04467 \text{ mmol NaBr}}{\text{mL}} \times 250.0 \text{ mL} \times \dfrac{102.9 \text{ g NaBr}}{1000 \text{ mmol}} \right)}{3.650 \text{ g sample}} \times 100\% = 31.48\% \text{ NaBr}$$

$$\frac{\left(\dfrac{0.04699 \text{ mmol NaBrO}_3}{\text{mL}} \times 250.0 \text{ mL} \times \dfrac{150.9 \text{ g NaBrO}_3}{1000 \text{ mmol}} \right)}{3.650 \text{ g sample}} \times 100\% = 48.57\% \text{ NaBrO}_3$$

17-21.

$$\text{Amnt EDTA reacted in 50.00 mL} = \left(\frac{0.05173 \text{ mmol EDTA}}{\text{mL}} \times 50.00 \text{ mL EDTA} \right) -$$

$$\left(\frac{0.06139 \text{ mmol Cu}^{2+}}{\text{mL}} \times 5.34 \text{ mL Cu}^{2+} \times \frac{1 \text{ mmol EDTA}}{\text{mmol Cu}^{2+}} \right) = 2.2587 \text{ mmol}$$

$$\text{Amnt EDTA reacted in 250.0 mL} = \text{Amnt (Ni+ Fe+ Cr)} = \frac{2.2587 \text{ mmol}}{\left(\dfrac{50.00 \text{ mL}}{250.0 \text{ mL}} \right)} = 11.2934 \text{ mmol}$$

$$\text{Amnt (Ni + Fe)} = \frac{\left(\dfrac{0.05173 \text{ mmol EDTA}}{\text{mL}} \times 36.98 \text{ mL EDTA} \right)}{\dfrac{50.00 \text{ mL}}{250.0 \text{ mL}}} = 9.5649 \text{ mmol}$$

$$\text{Amnt Cr} = 11.2934 \text{ mmol} - 9.5649 \text{ mmol} = 1.7285 \text{ mmol}$$

$$\text{Amnt Ni} = \frac{\left(\dfrac{0.05173 \text{ mmol EDTA}}{\text{mL}} \times 24.53 \text{ mL EDTA} \times \dfrac{1 \text{ mmol Ni}}{\text{mmol EDTA}} \right)}{\dfrac{50.00 \text{ mL}}{250.0 \text{ mL}}} = 6.3447 \text{ mmol}$$

$$\text{Amnt Fe} = 9.5649 \text{ mmol} - 6.3447 \text{ mmol} = 3.2202 \text{ mmol}$$

$$\%\text{Cr} = \frac{1.7285 \text{ mmol Cr} \times \dfrac{51.996 \text{ g Cr}}{1000 \text{ mmol}}}{0.6553 \text{ g}} \times 100\% = 13.72\%$$

$$\%\text{Ni} = \frac{6.3447 \text{ mmol Ni} \times \dfrac{58.69 \text{ g Ni}}{1000 \text{ mmol}}}{0.6553 \text{ g}} \times 100\% = 56.82\%$$

$$\%\text{Fe} = \frac{3.2202 \text{ mmol Fe} \times \dfrac{55.847 \text{ g Fe}}{1000 \text{ mmol}}}{0.6553 \text{ g}} \times 100\% = 27.44\%$$

17-22.

◢	A	B	C	D	E	F	G	H	I	J	K
1	Pb 17-38 Conditional constants for the Fe^{2+}-EDTA complex										
2	Note: The conditional constant K'_{MY} is the product of α_4 and K_{MY} (Equation 17-25).										
3	The value of K_{MY} is found in Table 17-4.										
4	K_{MY}	2.10E+14									
5	K_1	1.02E-02									
6	K_2	2.14E-03									
7	K_3	6.92E-07									
8	K_4	5.50E-11									
9	pH	D	α_4	K'_{MY}							
10	6.0	3.69E-17	2.25E-05	4.7E+09							
11	8.0	1.54E-19	5.39E-03	1.1E+12							
12	10.0	2.34E-21	3.55E-01	7.5E+13							
13	Spreadsheet Documentation										
14	Cell B10=(10^-A10)^4+$B$5*(10^-A10)^3+$B$5*$B$6*(10^-A10)^2+$B$5*$B$6*$B$7*(10^-A10)+$B$5*$B$6*$B$7*$B$8										
15	Cell C10=$B$5*$B$6*$B$7*$B$8/B10										
16	Cell D10=$B$4*C10										

17-23.

$$\text{Amnt Ca}^{2+} + \text{Amnt Mg}^{2+} = \left(\frac{0.01205 \text{ mmol EDTA}}{\text{mL}} \times 23.65 \text{ mL EDTA} \right) = 0.2850 \text{ mmol}$$

$$\text{Amnt Ca}^{2+} = \left(\frac{0.01205 \text{ mmol EDTA}}{\text{mL}} \times 14.53 \text{ mL EDTA} \times \frac{1 \text{ mmol Ca}^{2+}}{\text{mmol EDTA}} \right) = 0.1751 \text{ mmol}$$

$$\text{Amnt Mg}^{2+} = 0.2850 - 0.1751 = 0.1099 \text{ mmol}$$

(a)

See discussion of water hardness in 17D-9.

$$\text{Water hardness} \cong \text{Conc. CaCO}_3 \text{ in ppm} \approx \text{conc. Ca}^{2+} + \text{Mg}^{2+} \text{ in ppm}$$

$$= \frac{0.2850 \text{ mmol} \times \dfrac{100.087 \text{ mg CaCO}_3}{\text{mmol}}}{50.00 \text{ mL} \times \dfrac{L}{1000 \text{ mL}}} = 570.5 \text{ ppm CaCO}_3$$

(b)

$$\frac{\left(0.1751 \text{ mmol Ca}^{2+} \times \dfrac{1 \text{ mmol CaCO}_3}{\text{mmol Ca}^{2+}} \times \dfrac{100.08 \text{ mg CaCO}_3}{\text{mmol}} \right)}{50.00 \text{ mL} \times \dfrac{L}{1000 \text{ mL}}} = 350.5 \text{ ppm CaCO}_3$$

(c)

$$\frac{\left(0.1099 \text{ mmol Mg}^{2+} \times \dfrac{1 \text{ mmol MgCO}_3}{\text{mmol Mg}^{2+}} \times \dfrac{84.30 \text{ mg MgCO}_3}{\text{mmol}}\right)}{50.00 \text{ mL} \times \dfrac{\text{L}}{1000 \text{ mL}}} = 185.3 \text{ ppm MgCO}_3$$

Chapter 18

18-1. **(a)** *Oxidation* is a process in which a species loses one or more electrons.

   (c) A *salt bridge* provides electrical contact but prevents mixing of dissimilar solutions in an electrochemical cell.

   (e) The *Nernst equation* relates the potential to the concentrations (strictly, activities) of the participants in an electrochemical reaction.

18-2. **(a)** The *electrode potential* is the potential of an electrochemical cell in which a standard hydrogen electrode acts as the reference electrode on the left and the half-cell of interest is on the right.

   (c) The *standard electrode potential* is the potential of a cell consisting of the half-reaction of interest on the right and a standard hydrogen electrode on the left. The activities of all the participants in the half-reaction are specified as having a value of unity. The standard electrode potential is always a *reduction potential*.

18-3. **(a)** *Oxidation* is the process whereby a substance loses electrons; an *oxidizing agent* causes the loss of electrons.

   (c) The *cathode* of a cell is the electrode at which reduction occurs. The *right-hand electrode* is the electrode on the right in the cell diagram.

   (e) The *standard electrode potential* is the potential of a cell in which the standard hydrogen electrode acts as the reference electrode on the left and all participants in the right-hand electrode process have unit activity. The *formal potential* differs in that the molar *concentrations* of all the reactants and products are unity and the concentration of other species in the solution are carefully specified.

18-4. The first standard potential is for a solution saturated with I_2, which has an $I_2(aq)$ activity significantly less than one. The second potential is for a *hypothetical* half-cell in which the $I_2(aq)$ activity is unity.

18-5. To keep the solution saturated with $H_2(g)$. Only then is the hydrogen activity constant and the electrode potential constant and reproducible.

18-7. **(a)** $2Fe^{3+} + Sn^{2+} \rightarrow 2Fe^{2+} + Sn^{4+}$

(c) $2NO_3^- + Cu(s) + 4H^+ \rightarrow 2\,NO_2(g) + 2H_2O + Cu^{2+}$

(e) $Ti^{3+} + Fe(CN)_6^{3-} + H_2O \rightarrow TiO^{2+} + Fe(CN)_6^{4-} + 2H^+$

(g) $2Ag(s) + 2I^- + Sn^{4+} \rightarrow 2AgI(s) + Sn^2$

(i) $5HNO_2 + 2MnO_4^- + H^+ \rightarrow 5NO_3^- + 2Mn^{2+} + 3H_2O$

18-8. **(a)** Oxidizing agent Fe^{3+}; $Fe^{3+} + e^- \rightleftharpoons Fe^{2+}$

Reducing agent Sn^{2+}; $Sn^{2+} \rightleftharpoons Sn^{4+} + 2e^-$

(b) Oxidizing agent Ag^+; $Ag^+ + e^- \rightleftharpoons Ag(s)$

Reducing agent Cr; $Cr(s) \rightleftharpoons Cr^{3+} + 3e^-$

(c) Oxidizing agent NO_3^-, $NO_3^- + 2H^+ + e^- \rightleftharpoons NO_2(g) + H_2O$

Reducing agent Cu; $Cu(s) \rightleftharpoons Cu^{2+} + 2e^-$

(d) Oxidizing agent MnO_4^-; $MnO_4^- + 8H^+ + 5e^- \rightleftharpoons Mn^{2+} + 4H_2O$

Reducing agent H_2SO_3; $H_2SO_3 + H_2O \rightleftharpoons SO_4^{2-} + 4H^+ + 2e^-$

(e) Oxidizing agent $Fe(CN)_6^{3-}$; $Fe(CN)_6^{3-} + e^- \rightleftharpoons Fe(CN)_6^{4-}$

Reducing agent Ti^{3+}; Ti^{3+} + H$_2$O $\rightleftharpoons$ TiO^{2+} +2H$^+$ + e$^-$

(f) Oxidizing agent Ce^{4+}; Ce^{4+} + e$^-$ $\rightleftharpoons$ Ce^{3+}

Reducing agent H$_2$O$_2$; H$_2$O$_2$ $\rightleftharpoons$ O$_2$(g) + 2H$^+$ + 2e$^-$

(g) Oxidizing agent Sn^{4+}; Sn^{4+} + 2e$^-$ $\rightleftharpoons$ Sn^{2+}

Reducing agent Ag; Ag(s) + I$^-$ $\rightleftharpoons$ AgI(s) + e$^-$

(h) Oxidizing agent UO$_2^{2+}$; UO$_2^{2+}$ + 4H$^+$ + 2e$^-$ $\rightleftharpoons$ U^{4+} + 2H$_2$O

Reducing agent Zn; Zn(s) $\rightleftharpoons$ Zn^{2+} + 2e$^-$

(i) Oxidizing agent MnO$_4^-$; MnO$_4^-$ + 8H$^+$ + 5e$^-$ $\rightleftharpoons$ Mn^{2+} + 4H$_2$O

Reducing agent HNO$_2$ HNO$_2$ + H$_2$O $\rightleftharpoons$ NO$_3^-$ + 3H$^+$ + 2e$^-$

(j)) Oxidizing agent IO$_3^-$; IO$_3^-$ + 6H$^+$ + 2Cl$^-$ + 4e$^-$ $\rightleftharpoons$ ICl$_2^-$ + 3H$_2$O

Reducing agent H$_2$NNH$_2$; H$_2$NNH$_2$ $\rightleftharpoons$ N$_2$(g) + 4H$^+$ + 4e$^-$

18-9. **(a)** MnO$_4^-$ + 5VO^{2+} + 11H$_2$O $\rightarrow$ Mn^{2+} + 5V(OH)$_4^+$ + 2H$^+$

(c) Cr$_2$O$_7^{2-}$ + 3U^{4+} + 2H$^+$ $\rightarrow$ 2Cr^{3+} + 3UO$_2^{2+}$ + H$_2$O

(e) IO$_3^-$ + 5I$^-$ + 6H$^+$ $\rightarrow$ 3I$_2$ + 3H$_2$O

(g) HPO$_3^{2-}$ + 2MnO$_4^-$ + 3OH$^-$ $\rightarrow$ PO$_4^{3-}$ + 2MnO$_4^{2-}$ + 2H$_2$O

(i) V^{2+} + 2V(OH)$_4^+$ + 2H$^+$ $\rightarrow$ 3VO^{2+} + 5H$_2$O

18-11. (a) $AgBr(s) + e- \rightleftharpoons Ag(s) + Br-$ $V^{2+} \rightleftharpoons V^{3+} + e^-$

$Ti^{3+} + 2e- \rightleftharpoons Ti^+$ $Fe(CN)_6^{4-} \rightleftharpoons Fe(CN)_6^{3-} + e^-$

$V^{3+} + e- \rightleftharpoons V^{2+}$ $Zn \rightleftharpoons Zn^{2+} + 2e^-$

$Fe(CN)_6^{3-} + e^- \rightleftharpoons Fe(CN)_6^{4-}$ $Ag(s) + Br- \rightleftharpoons AgBr(s) + e^-$

$S_2O_8^{2-} + 2e^- \rightleftharpoons 2SO_4^{2-}$ $Ti+ \rightleftharpoons Ti^{3+} + 2e^-$

(b), (c)	E^0
$S_2O_8^{2-} + 2e^- \rightleftharpoons 2SO_4^{2-}$	2.01
$Ti^{3+} + 2e^- \rightleftharpoons Ti^+$	1.25
$Fe(CN)_6^{3-} + e^- \rightleftharpoons Fe(CN)_6^{4-}$	0.36
$AgBr(s) + e^- \rightleftharpoons Ag(s) + Br^-$	0.073
$V^{3+} + e^- \rightleftharpoons V^{2+}$	-0.256
$Zn^{2+} + 2e^- \rightleftharpoons Zn(s)$	-0.763

18-13. (a)

$$E_{Cu} = 0.337 - \frac{0.0592}{2}\log\left(\frac{1}{0.0380}\right) = 0.295 \text{ V}$$

(b)

$$K_{CuCl} = 1.9 \times 10^{-7} = [Cu^+][Cl^-]$$

$$E_{Cu} = 0.521 - \frac{0.0592}{1}\log\left(\frac{1}{[Cu^+]}\right) = 0.521 - \frac{0.0592}{1}\log\left(\frac{[Cl^-]}{K_{CuCl}}\right)$$

$$= 0.521 - \frac{0.0592}{1}\log\left(\frac{0.0650}{1.9 \times 10^{-7}}\right) = 0.521 - \frac{0.0592}{1}\log(3.42 \times 10^5)$$

$$= 0.521 - 0.328 = 0.193 \text{ V}$$

(c) $K_{Cu(OH)_2} = 4.8 \times 10^{-20} = [Cu^{2+}][OH^-]^2$

$$E_{Cu} = 0.337 - \frac{0.0592}{2}\log\left(\frac{1}{[Cu^{2+}]}\right) = 0.337 - \frac{0.0592}{2}\log\left(\frac{[OH^-]^2}{K_{Cu(OH)_2}}\right)$$

$$= 0.337 - \frac{0.0592}{2}\log\left(\frac{(0.0350)^2}{4.8 \times 10^{-20}}\right) = 0.337 - \frac{0.0592}{2}\log(2.55 \times 10^{16})$$

$$= 0.337 - 0.486 = -0.149 \text{ V}$$

(d) $\beta_4 = 5.62 \times 10^{11} = \dfrac{[Cu(NH_3)_4^{2+}]}{[Cu^{2+}][NH_3]^4}$

$$E_{Cu} = 0.337 - \frac{0.0592}{2}\log\left(\frac{1}{[Cu^{2+}]}\right) = 0.337 - \frac{0.0592}{2}\log\left(\frac{\beta_4[NH_3]^4}{[Cu(NH_3)_4^{2+}]}\right)$$

$$= 0.337 - \frac{0.0592}{2}\log\left(\frac{5.62 \times 10^{11}(0.108)^4}{0.0375}\right) = 0.337 - \frac{0.0592}{2}\log(2.04 \times 10^9)$$

$$= 0.337 - 0.276 = 0.061 \text{ V}$$

(e)

$$\frac{[CuY^{2-}]}{[Cu^{2+}]c_T} = \alpha_4 K_{CuY} = \left(3.6 \times 10^{-9}\right) \times \left(6.3 \times 10^{18}\right) = 2.3 \times 10^{10}$$

$$[CuY^{2-}] \approx 3.90 \times 10^{-3}$$

$$c_T = \left(3.90 \times 10^{-2}\right) - \left(3.90 \times 10^{-3}\right) = 0.0351$$

$$E_{Cu} = 0.337 - \frac{0.0592}{2}\log\left(\frac{1}{[Cu^{2+}]}\right) = 0.337 - \frac{0.0592}{2}\log\left(\frac{\alpha_4 K_{CuY^{2-}} c_T}{[CuY^{2-}]}\right)$$

$$= 0.337 - \frac{0.0592}{2}\log\left(\frac{2.3 \times 10^{10}(0.0351)}{3.90 \times 10^{-3}}\right) = 0.337 - \frac{0.0592}{2}\log(2.07 \times 10^{11})$$

$$= 0.337 - 0.335 = 0.002 \text{ V}$$

18-16. (a) $PtCl_4^{2-} + 2e^- \rightleftharpoons Pt(s) + 4 Cl^-$ $E^0 = 0.73$ V

$$E_{Pt} = 0.73 - \frac{0.0592}{2}\log\left(\frac{(0.2450)^4}{0.0160}\right) = 0.73 - (-0.019) = 0.75 \text{ V}$$

(b) $E^0 = 0.154$

$$E_{Pt} = 0.154 - \frac{0.0592}{2}\log\left(\frac{3.50\times10^{-3}}{6.50\times10^{-2}}\right) = 0.154 - (-0.038) = 0.192 \text{ V}$$

(c) pH = 6.50, $[H^+] = 3.16 \times 10^{-7}$

$$E_{Pt} = 0.000 - \frac{0.0592}{2}\log\left(\frac{1.00}{\left(3.16\times10^{-7}\right)^2}\right) = -0.385 \text{ V}$$

(d) $E^0 = 0.359$ V

$$E_{Pt} = 0.359 - \frac{0.0592}{1}\log\left(\frac{(0.0686)\times2}{(0.0255)\times(0.100)^2}\right) = 0.359 - 0.162 = 0.197 \text{ V}$$

(e) $2Fe^{3+} + Sn^{2+} \rightleftharpoons 2Fe^{2+} + Sn^{4+}$

$$\text{amount } Sn^{2+}\text{consumed} = \frac{0.0918 \text{ mmol } SnCl_2}{mL} \times \frac{1 \text{ mmol } Sn^{2+}}{\text{mmol } SnCl_2} \times 25.00 \text{ mL} = 2.295 \text{ mmol}$$

$$\text{amount } Fe^{3+}\text{consumed} = \frac{0.1568 \text{ mmol } FeCl_3}{mL} \times \frac{1 \text{ mmol } Fe^{3+}}{\text{mmol } FeCl_3} \times 25.00 \text{ mL} = 3.920 \text{ mmol}$$

$$\text{amount } Sn^{4+}\text{formed} = 3.920 \text{ mmol } Fe^{3+} \times \frac{1 \text{ mmol } Sn^{4+}}{2 \text{ mmol } Fe^{3+}} = 1.960 \text{ mmol}$$

$$\text{amount } Sn^{2+}\text{remaining} = 2.295 - 1.960 = 0.335 \text{ mmol}$$

$$E_{Pt} = 0.154 - \frac{0.0592}{2}\log\left(\frac{(0.335)/50.0}{(1.960)/50.0}\right) = 0.154 - (-0.023) = 0.177 \text{ V}$$

(f) $V(OH)_4^+ + V^{3+} + \rightleftharpoons 2VO^{2+} + 2H_2O$

$$V(OH)_4^+ + 2H^+ + 2e^- \rightleftharpoons VO^{2+} + 3H_2O \qquad E^0 = 1.00 \text{ V}$$

$$\text{amount} V(OH)_4^+\text{consumed} = \frac{0.0832 \text{ mmol } V(OH)_4^+}{mL} \times 25.00 \text{ mL} = 2.08 \text{ mmol}$$

$$\text{amount } V^{3+} \text{consumed} = \frac{0.01087 \text{ mmol } V_2(SO_4)_3}{mL} \times \frac{2 \text{ mmol } V^{3+}}{\text{mmol } V_2(SO_4)_3} \times 50.00 \text{ mL} = 1.087 \text{ mmol}$$

$$\text{amount } VO^{2+} \text{formed} = 1.087 \text{ mmol } V^{3+} \times \frac{2 \text{ mmol } VO^{2+}}{\text{mmol } V^{3+}} = 2.174 \text{ mmol}$$

$$\text{amount } V(OH)_4^+ \text{remaining} = 2.08 - 1.087 = 0.993 \text{ mmol}$$

$$E_{Pt} = 1.00 - 0.0592 \log\left(\frac{(2.174)/75.00}{(0.993/75.00)(0.1000)^2}\right) = 1.00 - 0.139 = 0.86 \text{ V}$$

18-18. (a)

$$E_{Ni} = -0.250 - \frac{0.0592}{2}\log\left(\frac{1.00}{0.0883}\right) = -0.250 - 0.031 = -0.281 \text{ V anode}$$

(b) $E_{Ag} = -0.151 - 0.0592 \log(0.0898) = -0.151 - (-0.062) = -0.089 \text{ V anode}$

(c)

$$E_{O_2} = 1.229 - \frac{0.0592}{4}\log\left(\frac{1.00}{(780/760)(2.50\times10^{-4})^4}\right) = 1.229 - 0.213 = 1.016 \text{ V cathode}$$

(d) $E_{Pt} = 0.154 - \frac{0.0592}{2}\log\left(\frac{0.0893}{0.215}\right) = 0.154 - (-0.011) = 0.165 \text{ V cathode}$

(e) $E_{Ag} = 0.017 - 0.0592 \log\left(\frac{(0.1035)^2}{0.00891}\right) = 0.017 - 0.005 = 0.012 \text{ V cathode}$

18-20. $2Ag^+ + 2e^- \rightleftharpoons 2Ag(s) \qquad E^o = 0.779$

$$[Ag^+]^2[SO_3^{2-}] = 1.5 \times 10^{-14} = K_{sp}$$

$$E = 0.799 - \frac{0.0592}{2}\log\left(\frac{1}{[Ag^+]^2}\right) = 0.799 - \frac{0.0592}{2}\log\left(\frac{[SO_3^{2-}]}{K_{sp}}\right)$$

When $[SO_3^{2-}] = 1.00$, $E = E^o$ for $Ag_2SO_3(s) + 2e^- \rightleftharpoons 2Ag(s) + SO_3^{2-}$.

Thus,

$$E = 0.799 - \frac{0.0592}{2}\log\left(\frac{1.00}{K_{sp}}\right) = 0.799 - \frac{0.0592}{2}\log\left(\frac{1.00}{1.5\times10^{-14}}\right) = 0.799 - 0.409 = 0.390 \text{ V}$$

18-22. $2Tl^+ + 2e^- \rightleftharpoons 2Tl(s)$ $E^\circ = -0.336$

$$[Tl^+]^2[S^{2-}] = 6\times10^{-22} = K_{sp}$$

$$E = -0.336 - \frac{0.0592}{2}\log\left(\frac{1}{[Tl^+]^2}\right) = -0.336 - \frac{0.0592}{2}\log\left(\frac{[S^{2-}]}{K_{sp}}\right)$$

When $[S^{2-}] = 1.00$, $E = E^\circ$ for $Tl_2S(s) + 2e^- \rightleftharpoons 2Tl(s) + S^{2-}$.

Thus,

$$E = -0.336 - \frac{0.0592}{2}\log\left(\frac{1.00}{K_{sp}}\right) = -0.336 - \frac{0.0592}{2}\log\left(\frac{1.00}{6\times10^{-22}}\right)$$

$$= -0.336 - 0.628 = -0.96 \text{ V}$$

18-24. $E = -0.763 - \frac{0.0592}{2}\log\left(\frac{1}{[Zn^{2+}]}\right)$

$$\frac{[ZnY^{2-}]}{[Zn^{2+}][Y^{4-}]} = 3.2\times10^{16}$$

$$E = -0.763 - \frac{0.0592}{2}\log\left(\frac{[Y^{4-}](3.2\times10^{16})}{[ZnY^{2-}]}\right)$$

When $[Y^{4-}] = [ZnY^{2-}] = 1.00$, $E = E^\circ_{ZnY^{2-}}$

$$E = -0.763 - \frac{0.0592}{2}\log\left(\frac{1.00(3.2\times10^{16})}{1.00}\right) = -0.763 - 0.489 = -1.25 \text{ V}$$

18-25. $[Fe^{3+}] = \dfrac{[FeY^-]}{[Y^{4-}](1.3\times10^{25})}$ and $[Fe^{2+}] = \dfrac{[FeY^{2-}]}{[Y^{4-}](2.1\times10^{14})}$

$$E = 0.771 - 0.0592 \log\left(\frac{[Fe^{2+}]}{[Fe^{3+}]}\right)$$

$$= 0.771 - 0.0592 \log\left(\frac{[FeY^{2-}](1.3\times10^{25})}{[FeY^-](2.1\times10^{14})}\right)$$

When $[FeY^{2-}] = [FeY^-] = 1.00$, $E = E^{\circ}_{FeY^-}$

$$E = 0.771 - 0.0592 \log\left(\frac{1.00(1.3\times10^{25})}{1.00(2.1\times10^{14})}\right) = 0.771 - 0.64 = 0.13 \text{ V}$$

Chapter 19

19-1. The electrode potential of a system that contains two or more redox couples is the electrode potential of all half-cell processes at equilibrium in the system.

19-2. **(a)** *Equilibrium* is the state that a system assumes after each addition of reagent. *Equivalence* refers to a particular equilibrium state when a stoichiometric amount of titrant has been added.

19-4. For points before the equivalence point, potential data are computed from the analyte standard potential and the analytical concentrations of the analyte and its reaction product. Post-equivalence point data are based upon the standard potential for the titrant and its analytical concentrations. The equivalence point potential is computed from the two standard potentials and the stoichiometric relation between the analyte and titrant.

19-6. An asymmetric titration curve will be encountered whenever the titrant and the analyte react in a ratio that is not 1:1.

19-8. **(a)**

$$E_{right} = -0.277 - \frac{0.0592}{2} \log\left(\frac{1}{5.87 \times 10^{-4}}\right) = -0.373 \text{ V}$$

$$E_{left} = -0.763 - \frac{0.0592}{2} \log\left(\frac{1}{0.100}\right) = -0.793 \text{ V}$$

$$E_{cell} = E_{right} - E_{left} = -0.373 - (-0.793) = 0.420 \text{ V}$$

Because E_{cell} is positive, the reaction would proceed spontaneously in the direction considered (oxidation on the left, reduction on the right).

(b) $$E_{right} = 0.854 - \frac{0.0592}{2} \log\left(\frac{1}{0.0350}\right) = 0.811 \text{ V}$$

$$E_{left} = 0.771 - \frac{0.0592}{1} \log\left(\frac{0.0700}{0.1600}\right) = 0.792 \text{ V}$$

150

$$E_{cell} = E_{right} - E_{left} = 0.811 - 0.792 = 0.019 \text{ V}$$

Because E_{cell} is positive, the spontaneous reaction would be oxidation on the left and reduction on the right.

(c)

$$E_{right} = 1.229 - \frac{0.0592}{4} \log\left(\frac{1}{1.12(0.0333)^4}\right) = 1.142 \text{ V}$$

$$E_{left} = 0.799 - 0.0592 \log\left(\frac{1}{0.0575}\right) = 0.726 \text{ } V$$

$$E_{cell} = E_{right} - E_{left} = 1.142 - 0.726 = 0.416 \text{ V}$$

The spontaneous reaction would be oxidation on the left, reduction on the right.

(d)

$$E_{right} = -0.151 - 0.0592 \log(0.1220) = -0.097 \text{ V}$$

$$E_{left} = 0.337 - \frac{0.0592}{2} \log\left(\frac{1}{0.0420}\right) = 0.296 \text{ V}$$

$$E_{cell} = E_{right} - E_{left} = -0.097 - 0.296 = -0.393 \text{ V}$$

The spontaneous reaction would be reduction on the left, oxidation on the right, not the reaction in the direction considered.

(e)

$$\frac{[H_3O^+][HCOO^-]}{[HCOOH]} = 1.80 \times 10^{-4} = \frac{[H_3O^+]0.0700}{0.1400}$$

$$[H_3O^+] = \frac{(1.80 \times 10^{-4})(0.1400)}{0.0700} = 3.60 \times 10^{-4}$$

$$E_{right} = 0.000 - \frac{0.0592}{2} \log \left(\frac{1.00}{\left(3.60 \times 10^{-4} \right)^2} \right) = -0.204 \text{ V}$$

$$E_{left} = 0.000 \ V$$

$$E_{cell} = -0.204 - 0.000 = -0.204 \text{ V}$$

Because E_{cell} is negative, the reaction woulds not proceed spontaneously in the direction considered (reduction on the left, oxidation on the right).

(f)

$$E_{right} = 0.771 - 0.0592 \log \left(\frac{0.1134}{0.003876} \right) = 0.684 \text{ V}$$

$$E_{left} = 0.334 - \frac{0.0592}{2} \log \left(\frac{4.00 \times 10^{-2}}{\left(8.00 \times 10^{-3} \right) \left(1.00 \times 10^{-3} \right)^4} \right) = -0.042 \ V$$

$$E_{cell} = E_{right} - E_{left} = 0.684 - (-0.042) = 0.726 \text{ V}$$

The direction considered (oxidation on the left, reduction on the right) is the spontaneous direction.

19-9. (a)

$$E_{Pb^{2+}} = -0.126 - \frac{0.0592}{2} \log \left(\frac{1}{0.0220} \right) = -0.175 \text{ V}$$

$$E_{Zn^{2+}} = -0.763 - \frac{0.0592}{2} \log \left(\frac{1}{0.1200} \right) = -0.790 \text{ V}$$

$$E_{cell} = E_{right} - E_{left} = -0.175 - (-0.790) = 0.615 \text{ V}$$

(c)

$$E_{SHE} = 0.000 \ V$$

$$E_{TiO^{2+}} = 0.099 - 0.0592 \log \left(\frac{0.07000}{\left(3.50 \times 10^{-3} \right) \left(10^{-3} \right)^2} \right) = -0.333 \text{ V}$$

$$E_{cell} = E_{right} - E_{left} = -0.333 - 0.000 = -0.333 \text{ V}$$

19-11. Note that in these calculations, it is necessary to round the answers to either one or two significant figures because the final step involves taking the antilogarithm of a large number.

(a) $Fe^{3+} + V^{2+} \rightleftharpoons Fe^{2+} + V^{3+}$ $E^o_{Fe^{3+}} = 0.771$ $E^o_{V^{3+}} = -0.256$

$$0.771 - 0.0592 \log\left(\frac{[Fe^{2+}]}{[Fe^{3+}]}\right) = -0.256 - 0.0592 \log\left(\frac{[V^{2+}]}{[V^{3+}]}\right)$$

$$\frac{0.771 - (-0.256)}{0.0592} = \log\left(\frac{[Fe^{2+}][V^{3+}]}{[Fe^{3+}][V^{2+}]}\right) = \log K_{eq} = 17.348$$

$$K_{eq} = 2.23 \times 10^{17} \; (2.2 \times 10^{17})$$

(c) $2V(OH)_4^+ + U^{4+} \rightleftharpoons 2VO^{2+} + UO_2^{2+} + 4H_2O$ $E^o_{V(OH)_4^+} = 1.00$ $E^o_{UO_2^{2+}} = 0.334$

$$1.00 - \frac{0.0592}{2}\log\left(\frac{[VO^{2+}]^2}{[V(OH)_4^+]^2[H^+]^4}\right) = 0.334 - \frac{0.0592}{2}\log\left(\frac{[U^{4+}]}{[UO_2^{2+}][H^+]^4}\right)$$

$$\frac{(1.00 - 0.334)\,2}{0.0592} = \log\left(\frac{[VO^{2+}]^2[UO_2^{2+}]}{[V(OH)_4^+]^2[U^{4+}]}\right) = \log K_{eq} = 22.50$$

$$K_{eq} = 3.2 \times 10^{22} \; (3 \times 10^{22})$$

(e)

$$2Ce^{4+} + H_3AsO_3 + H_2O \rightleftharpoons 2Ce^{3+} + H_3AsO_4 + 2H^+$$

$$E^o_{Ce^{4+}} (\text{in 1 M } HClO_4) = 1.70 \quad E^o_{H_3AsO_4} = 0.577$$

$$1.70 - \frac{0.0592}{2}\log\left(\frac{[Ce^{3+}]^2}{[Ce^{4+}]^2}\right) = 0.577 - \frac{0.0592}{2}\log\left(\frac{[H_3AsO_4]}{[H_3AsO_3][H^+]^2}\right)$$

$$\frac{(1.70 - 0.577)\,2}{0.0592} = \log\left(\frac{[Ce^{3+}]^2[H_3AsO_3][H^+]^2}{[Ce^{4+}]^2[H_3AsO_4]}\right) = \log K_{eq} = 37.94$$

$$K_{eq} = 8.9 \times 10^{37} \; (9 \times 10^{37})$$

(g) $VO^{2+} + V^{2+} + 2H^+ \rightleftharpoons 2V^{3+} + H_2O$ $E^o_{VO^{2+}} = 0.359$ $E^o_{V^{3+}} = -0.256$

$$0.359 - 0.0592 \log\left(\frac{[V^{3+}]}{[VO^{2+}][H^+]^2}\right) = -0.256 - 0.0592 \log\left(\frac{[V^{2+}]}{[V^{3+}]}\right)$$

$$\frac{0.359 - (-0.256)}{0.0592} = \log\left(\frac{[V^{3+}]^2}{[VO^{2+}][H^+]^2[V^{2+}]}\right) = \log K_{eq} = 10.389$$

$$K_{eq} = 2.4 \times 10^{10}$$

19-14.

	E_{eq}, V	Indicator
(a)	0.258	Phenosafranine
(b)	−0.024	None
(c)	0.444	Indigo tetrasulfonate or Methylene blue
(d)	1.09	1,10-Phenanthroline
(e)	0.951	Erioglaucin A
(f)	0.330	Indigo tetrasulfonate
(g)	−0.008	None
(h)	−0.194	None

Chapter 20

20-1. **(a)** $2Mn^{2+} + 5S_2O_8^{2-} + 8H_2O \rightarrow 10SO_4^{2-} + 2MnO_4^- + 16H^+$

(c) $H_2O_2 + U^{4+} \rightarrow UO_2^{2+} + 2H^+$

(e) $2MnO_4^- + 5H_2O_2 + 6H^+ \rightarrow 5O_2 + 2Mn^{2+} + 8H_2O$

20-2. Only in the presence of Cl^- ion is Ag a sufficiently good reducing agent to be very useful for prereductions. In the presence of Cl^-, the half-reaction occurring in the Walden reductor is

$$Ag(s) + Cl^- \rightarrow AgCl(s) + e^-$$

The excess HCl increases the tendency of this reaction to occur by the common ion effect.

20-4. Standard solutions of reductants find somewhat limited use because of their susceptibility to air oxidation.

20-6. Freshly prepared solutions of permanganate are inevitably contaminated with small amounts of solid manganese dioxide, which catalyzes the further decompositions of permanganate ion. By removing the dioxide at the outset, a much more stable standard reagent is produced.

20-8. Solutions of $K_2Cr_2O_7$ are used extensively for back-titrating solutions of Fe^{2+} when the latter is being used as a standard reductant for the determination of oxidizing agents.

20-10. When a measured volume of a standard solution of KIO_3 is introduced into an acidic solution containing an excess of iodide ion, a known amount of iodine is produced as a result of:

$$IO_3^- + 5I^- + 6H^+ \rightarrow 3I_2 + 3H_2O$$

155

20-12. Starch decomposes in the presence of high concentrations of iodine to give products that do not behave satisfactorily as indicators. This reaction is prevented by delaying the addition of the starch until the iodine concentration is very small.

20-13. $0.2541 \text{ g sample} \times \dfrac{1000 \text{ mmol Fe}^{2+}}{55.847 \text{ g}} = 4.5499 \text{ mmol Fe}^{2+}$

(a) $\dfrac{4.5499 \text{ mmol Fe}^{2+}}{36.76 \text{ mL}} \times \dfrac{1 \text{ mmol Ce}^{4+}}{\text{mmol Fe}^{2+}} = 0.1238 \text{ M Ce}^{4+}$

(c) $\dfrac{4.5499 \text{ mmol Fe}^{2+}}{36.76 \text{ mL}} \times \dfrac{1 \text{ mmol MnO}_4^{-}}{5 \text{ mmol Fe}^{2+}} = 0.02475 \text{ M MnO}_4^{-}$

(e) $\dfrac{4.5499 \text{ mmol Fe}^{2+}}{36.76 \text{ mL}} \times \dfrac{1 \text{ mmol IO}_3^{-}}{4 \text{ mmol Fe}^{2+}} = 0.03094 \text{ M IO}_3^{-}$

20-14. $\dfrac{0.05000 \text{ mol KBrO}_3}{\text{L}} \times 1.000 \text{ L} \times \dfrac{167.001 \text{ g KBrO}_3}{\text{mol}} = 8.350 \text{ g KBrO}_3$

Dissolve 8.350 g $KBrO_3$ in water and dilute to 1.000 L.

20-16. $Ce^{4+} + Fe^{2+} \rightarrow Ce^{3+} + Fe^{3+}$

$\dfrac{0.2219 \text{ g}}{34.65 \text{ mL Ce}^{4+}} \times \dfrac{1000 \text{ mL}}{\text{L}} \times \dfrac{1 \text{ mol Fe}}{55.847 \text{ g}} \times \dfrac{1 \text{ mol Fe}^{2+}}{\text{mol Fe}} \times \dfrac{1 \text{ mol Ce}^{4+}}{\text{mol Fe}^{2+}} = 0.1147 \text{ M Ce}^{4+}$

20-18. $MnO_2 + 2I^- + 4H^+ \rightarrow Mn^{2+} + I_2 + 2H_2O$

$I_2 + 2S_2O_3^{2-} \rightarrow 2I^- + S_4O_6^{2-}$

$1 \text{ mmol MnO}_2 = 1 \text{ mmol I}_2 = 2 \text{ mmol S}_2O_3^{2-}$

$$\dfrac{\left(\dfrac{0.08041 \text{ mmol}}{\text{mL}} \times 29.62 \text{ mL Na}_2S_2O_3 \times \dfrac{1 \text{ mmol MnO}_2}{2 \text{ mmol Na}_2S_2O_3} \times \dfrac{86.937 \text{ g MnO}_2}{1000 \text{ mmol}} \right)}{0.1267 \text{ g sample}} \times 100\%$$

$= 81.71\% \text{ MnO}_2$

20-20.

$$2H_2NOH + 4Fe^{3+} \rightleftharpoons N_2O(g) + 4Fe^{2+} + 4H^+ + H_2O$$

$$Cr_2O_7^{2-} + 6Fe^{2+} + 14H^+ \rightleftharpoons 2Cr^{3+} + 6Fe^{3+} + 7H_2O$$

1 mmol $Cr_2O_7^{2-}$ = 6 mmol Fe^{3+} = 3 mmol H_2NOH

$$\frac{\left(\dfrac{0.01528 \text{ mmol } K_2Cr_2O_7}{\text{mL}} \times 14.48 \text{ mL } K_2Cr_2O_7 \times \dfrac{3 \text{ mmol } H_2NOH}{\text{mmol } K_2Cr_2O_7} \right)}{25.00 \text{ mL sample}}$$

$$= 0.0266 \text{ M } H_2NOH$$

20-22.　　$H_3AsO_3 + I_2 + H_2O \rightarrow H_3AsO_4 + 2I^- + 2H^+$

1 mmol I_2 = 1 mmol H_3AsO_3 = ½ mmol As_2O_3

$$\frac{\left(\dfrac{0.03142 \text{ mmol } I_2}{\text{mL}} \times 31.36 \text{ mL } I_2 \times \dfrac{1 \text{ mmol } As_2O_3}{2 \text{ mmol } I_2} \times \dfrac{197.841 \text{ g } As_2O_3}{1000 \text{ mmol}} \right)}{8.13 \text{ g sample}} \times 100\%$$

$$= 1.199\% \text{ As}_2O_3$$

20-24.

$$2I^- + Br_2 \rightarrow I_2 + 2Br^-$$

$$IO_3^- + 5I^- + 6H^+ \rightarrow 3I_2 + 3H_2O$$

$$I_2 + 2S_2O_3^{2-} \rightarrow 2I^- + S_4O_6^{2-}$$

1 mmol KI = 1 mmol IO_3^- = 3 mmol I_2 = 6 mmol $S_2O_3^{2-}$

$$\frac{\left(\dfrac{0.04926 \text{ mmol } Na_2S_2O_3}{\text{mL}} \times 19.72 \text{ mL } Na_2S_2O_3 \times \dfrac{1 \text{ mmol KI}}{6 \text{ mmol } Na_2S_2O_3} \times \dfrac{166.00 \text{ g KI}}{1000 \text{ mmol}} \right)}{1.307 \text{ g sample}} \times 100\%$$

$$= 2.056\% \text{ KI}$$

20-26.

$$SO_2(g) + 2OH^- \rightarrow SO_3^{2-} + H_2O$$

$$IO_3^- + 2H_2SO_3 + 2Cl^- \rightarrow ICl_2^- + 2SO_4^{2-} + 2H^+ + H_2O$$

$$1 \text{ mmol } IO_3^- = 2 \text{ mmol } H_2SO_3 = 2 \text{ mmol } SO_2$$

In $\dfrac{2.50 \text{ L}}{\text{min}} \times 59.00 \text{ min} = 147.5 \text{ L of sample, there are}$

$$\frac{0.002997 \text{ mmol KIO}_3}{\text{mL}} \times 5.15 \text{ mL KIO}_3 \times \frac{2 \text{ mmol SO}_2}{\text{mmol KIO}_3} \times \frac{64.065 \text{ g SO}_2}{1000 \text{ mmol}} = 1.9776 \times 10^{-3} \text{ g SO}_2$$

$$\left(\frac{1.9776 \times 10^{-3} \text{ g SO}_2}{147.5 \text{ L} \times \dfrac{1.20 \text{ g}}{\text{L}}} \right) \times 10^6 \text{ ppm}$$

$$= 11.2 \text{ ppm SO}_2$$

20-28.

$$O_2 + 4Mn(OH)_2(s) + 2H_2O \rightleftharpoons 4Mn(OH)_3(s)$$

$$4Mn(OH)_3(s) + 12H^+ + 4I^- \rightleftharpoons 4Mn^{2+} + 2I_2 + 6H_2O$$

$$\frac{0.00897 \text{ mmol S}_2O_3^{2-}}{\text{mL}} \times 14.60 \text{ mL S}_2O_3^{2-} \times \frac{1 \text{ mmol O}_2}{4 \text{ mmol S}_2O_3^{2-}} \times \frac{32.0 \text{ mg O}_2}{\text{mmol}} = 1.0477 \text{ mg O}_2$$

$$\frac{1.0477 \text{ mg O}_2}{\left(25 \text{ mL sample} \times \dfrac{250 \text{ mL}}{254 \text{ mL}} \right)}$$

$$= 0.0426 \text{ mg O}_2/\text{mL sample}$$

Chapter 21

21-1.   **(a)** An *indicator electrode* is an electrode used in potentiometry that responds to

variations in the activity of an analyte ion or molecule.

(c) An *electrode of the first kind* is a metal electrode that responds to the activity of its

cation in solution.

21-2.   **(a)** A *liquid junction potential* is the potential that develops across the interface between

two solutions having different electrolyte compositions.

(c) The *asymmetry potential* is a potential that develops across an ion-sensitive membrane

when the concentrations of the ion are the same on either side of the membrane. This

potential arises from dissimilarities between the inner and outer surface of the membrane.

21-3.   **(a)** A titration is generally more accurate than measurements of electrode potential.

Therefore, if ppt accuracy is needed, a titration should be picked.

(b) Electrode potentials are related to the activity of the analyte. Thus, pick potential

measurements if activity is the desired quantity.

21-5.   The potential arises from the difference in positions of dissociation equilibria on each of

the two surfaces. These equilibria are described by

$$\underset{\text{membrane}}{H^+Gl^-} \rightleftharpoons \underset{\text{solution}}{H^+} + \underset{\text{membrane}}{Gl^-}$$

The surface exposed to the solution having the higher H^+ concentration becomes positive

with respect to the other surface. This charge difference, or potential, serves as the

analytical parameter when the pH of the solution on one side of the membrane is held

constant.

21-7. Uncertainties include (1) the acid error in highly acidic solutions, (2) the alkaline error in strongly basic solutions, (3) the error that arises when the ionic strength of the calibration standards differs from that of the analyte solution, (4) uncertainties in the pH of the standard buffers, (5) nonreproducible junction potentials with solutions of low ionic strength and (6) dehydration of the working surface.

21-9. The *alkaline error* arises when a glass electrode is employed to measure the pH of solutions having pH values in the 10 to 12 range or greater. In the presence of alkali ions, the glass surface becomes responsive to not only hydrogen ions but also alkali metal ions. Measured pH values are low as a result.

21-11. (b) The *boundary potential* for a membrane electrode is a potential that develops when the membrane separates two solutions that have different concentrations of a cation or an anion that the membrane binds selectively. For an aqueous solution, the following equilibria develop when the membrane is positioned between two solutions of A^+:

$$\underset{\text{membrane}_1}{A^+M^-} \overset{\rightarrow}{\leftarrow} \underset{\text{solution}_1}{A^+} + \underset{\text{membrane}_1}{M^-}$$

$$\underset{\text{membrane}_2}{A^+M^-} \overset{\rightarrow}{\leftarrow} \underset{\text{solution}_2}{A^+} + \underset{\text{membrane}_2}{M^-}$$

where the subscripts refer to the two sides of the membrane. A potential develops across this membrane if one of these equilibria proceeds further to the right than the other, and this potential is the boundary potential. For example, if the concentration of A^+ is greater in solution 1 than in solution 2, the negative charge on side 1 of the membrane will be less than that of side 2 because the equilibrium on side 1 will lie further to the left. Thus, a greater fraction of the negative charge on side 1 will be neutralized by A^+.

(d) The membrane in a solid-state electrode for F⁻ is crystalline LaF_3, which when immersed in aqueous solution, dissociates according to the equation

$$LaF_3(s) \rightleftharpoons La^{3+} + 3F^-$$

Thus, the boundary potential develops across this membrane when it separates two solutions of F⁻ ion concentration. The source of this potential is the same as described in part (b).

21-12. The direct potentiometric measurement of pH provides a measure of the equilibrium activity of hydronium ions in the sample. A potentiometric titration provides information on the amount of reactive protons, both ionized and nonionized, in the sample.

21-15. $AgIO_3(s) + e^- \rightleftharpoons Ag(s) + IO_3^-$

(a)

$$E_{Ag} = 0.799 - 0.0592 \log\left(\frac{1}{[Ag^+]}\right) \quad K_{sp} = [Ag^+][IO_3^-] = 3.1\times10^{-8}$$

$$E_{Ag} = 0.799 - 0.0592 \log\left(\frac{[IO_3^-]}{K_{sp}}\right)$$

When $[IO_3^-] = 1.00$, E_{Ag} is equal to $E^o_{AgIO_3}$ for the reduction of $AgIO_3$, that is,

$$E^o_{AgIO_3} = 0.799 - 0.0592 \log\left(\frac{1.00}{3.1\times10^{-8}}\right) = 0.354 \text{ V}$$

(b) $SCE\|IO_3^- (x\ M), AgIO_3(sat'd)\,|\,Ag$

(c)

$$E_{cell} = E_{AgIO_3} - E_{SCE}$$

$$= \left(0.354 - 0.0592\log\left([IO_3^-]\right) - 0.244\right)$$

$$= 0.110 + 0.0592\,pIO_3$$

$$pIO_3 = \frac{E_{cell} - 0.110}{0.0592}$$

(d) $pIO_3 = \dfrac{0.306 - 0.110}{0.0592} = 3.31$

21-17. (a) SCE$\|$I$^-$ (x M), AgI (sat'd) $|$ Ag

(c) SCE$\|$PO$_4^{3-}$ (x M), Ag$_3$PO$_4$ (sat'd) $|$ Ag

21-19. (a) $pI = \dfrac{-0.196 + 0.395}{0.0592} = 3.36$

(c) $pPO_4 = \dfrac{3\left(0.211 - 0.163\right)}{0.0592} = 2.43$

21-20. SCE$\|$Ag$_2$CrO$_4$ (sat'd), CrO$_4^{2-}$ (x M) $|$ Ag

$$Ag_2CrO_4(s) + 2e^- \rightleftharpoons 2Ag(s) + CrO_4^{2-} \qquad E^o = 0.446\ V$$

$$0.336 = 0.446 - \frac{0.0592}{2}\log\left([CrO_4^{2-}]\right) - 0.244 = 0.202 + \frac{0.0592}{2}pCrO_4$$

$$pCrO_4 = \frac{2\left(0.389 - 0.202\right)}{0.0592}$$

$$pCrO_4 = 6.32$$

21-21. Substituting into Equation 21-22 gives

$$pH = -\frac{1(E_{cell} - K)}{0.0592} \quad \text{and} \quad 4.006 = -\frac{(0.2106 - K)}{0.0592}$$

$$K = (4.006 \times 0.0592) + 0.2106 = 0.447755$$

(a) $pH = -\dfrac{(-0.2902 - 0.447755)}{0.0592} = 12.47$

$$a_{H^+} = \text{antilog}(-12.4655) = 3.42 \times 10^{-13} \text{ M}$$

(b) $pH = -\dfrac{(0.1241 - 0.447755)}{0.0592} = 5.47$

$$a_{H^+} = \text{antilog}(-5.4671) = 3.41 \times 10^{-6} \text{ M}$$

(c) For part (a)

If $E = -0.2902 + 0.002 = -0.2882$ V

$$pH = -\frac{(-0.2882 - 0.447755)}{0.0592} = 12.43$$

$$a_{H^+} = \text{antilog}(-12.4317) = 3.70 \times 10^{-13}$$

If $E = -0.2902 - 0.002 = -0.2922$ V

$$pH = -\frac{(-0.2922 - 0.447755)}{0.0592} = 12.50$$

$$a_{H^+} = \text{antilog}(-12.4992) = 3.17 \times 10^{-13} \text{ M}$$

Thus pH should be 12.43 to 12.50 and a_{H^+} in the range of 3.17 to 3.70×10^{-13} M

Proceeding in the same way for (b), we obtain

pH in the range 5.43 to 5.50

a_{H^+} in the range 3.16×10^{-6} to 3.69×10^{-6} M

21-22.

$$\text{amount HA} = \frac{0.1243 \text{ mmol NaOH}}{\text{mL}} \times 18.62 \text{ mL NaOH} \times \frac{1 \text{ mmol HA}}{\text{mmol NaOH}} = 2.3145 \text{ mmol}$$

$$\frac{0.4021 \text{ g HA}}{2.3145 \text{ mmol HA}} \times \frac{1000 \text{ mmol}}{\text{mol}} = \frac{173.7 \text{ g HA}}{\text{mol}}$$

$$\mathcal{M}_{HA} = 173.7 \text{ g/mol}$$

21-26.

$$pNa = -\log\left([Na^+]\right) = -\left(\frac{E'_{cell} - K}{0.0592}\right) \text{ where } E'_{cell} = -0.2462 \text{ V}$$

After addition $E''_{cell} = -0.1994$ V

$$-\log\left(\frac{10.00 \times [Na^+] + 1.00 \times (2.00 \times 10^{-2})}{10.00 + 1.00}\right) = -\left(\frac{E''_{cell} - K}{0.0592}\right)$$

$$-\log\left(0.9091[Na^+] + (1.818 \times 10^{-3})\right) = -\left(\frac{E''_{cell} - K}{0.0592}\right)$$

Subtracting this latter equation from that for the initial potential gives

$$-\log\left([Na^+]\right) + \log\left(0.9091[Na^+] + (1.818 \times 10^{-3})\right) = -\left(\frac{E'_{cell} - K}{0.0592}\right) + \left(\frac{E''_{cell} - K}{0.0592}\right)$$

$$= \left(\frac{E''_{cell} - E'_{cell}}{0.0592}\right)$$

$$-\log\left(\frac{[Na^+]}{0.9091[Na^+] + (1.818 \times 10^{-3})}\right) = \frac{-0.1994 + 0.2462}{0.0592} = 0.7905$$

$$\text{or, } \log\left(\frac{[Na^+]}{0.9091[Na^+] + (1.818 \times 10^{-3})}\right) = -0.7905$$

$$\frac{[Na^+]}{0.9091[Na^+] + (1.818 \times 10^{-3})} = \text{antilog}(-0.7905) = 0.16198$$

$$[Na^+] = 0.1473[Na^+] + 2.945 \times 10^{-4}$$

$$[Na^+] = 3.453 \times 10^{-4} \text{ M or rounding } 3.5 \times 10^{-4} \text{ M}$$

Chapter 22

22-1. **(a)** In *Concentration polarization*, the current in an electrochemical cell is limited by the

rate at which reactants are brought to or removed from the surface of one or both

electrodes.   In *Kinetic polarization*, the current is limited by the rate at which electrons

are transferred between the electrode surfaces and the reactant in solution. For either

type, the current is no longer linearly related to cell potential.

(c) *Diffusion* is the movement of species under the influence of a concentration gradient.

Migration is the movement of an ion under the influence of an electrostatic attractive or

repulsive force.

(e) The *electrolysis circuit* consists of a working electrode and a counter electrode. The

control circuit regulates the applied potential such that the potential between the working

electrode and a reference electrode in the control circuit is constant and at a desired level.

22-2. **(a)** *Ohmic potential*, or *IR* drop, of a cell is the product of the current in the cell in

amperes and the electrical resistance of the cell in ohms.

(c) In *controlled-potential electrolysis*, the potential applied to a cell is continuously

adjusted to maintain a constant potential between the working electrode and a reference

electrode.

(e) *Current efficiency* is a measure of agreement between the number of faradays of

charge and the number of moles of reactant oxidized or reduced at a working electrode.

22-3. *Diffusion* arises from concentration differences between the electrode surface and the

bulk of solution. *Migration* results from electrostatic attraction or repulsion. *Convection*

results from stirring, vibration or temperature differences.

22-5. Variables that influence concentration polarization include temperature, stirring, reactant concentrations, presence or absence of other electrolytes and electrode surface areas.

22-7. Kinetic polarization is often encountered when the product of a reaction is a gas, particularly when the electrode is a soft metal such as mercury, zinc or copper. It is likely to occur at low temperatures and high current densities.

22-9. Potentiometric methods are carried out under zero current conditions and the effect of the measurement on analyte concentration is typically undetectable. In contrast, electrogravimetric and coulometric methods depend on the presence of a net current and a net cell reaction (i.e., the analyte is quantitatively converted to a new oxidation state). Unlike potentiometric methods where the cell potential is simply the difference between two electrode potentials, two additional phenomena, *IR* drop and polarization, must be considered in electrogravimetric and coulometric methods where current is present. Finally, the final measurement in electrogravimetric and coulometric methods is the mass of the product produced electrolytically, while in potentiometric methods it is the cell potential.

22-11. The species produced at the counter electrode are potential interferences by reacting with the products at the working electrode. Isolation of one from the other is ordinarily required.

22-13.

(b)

$$\frac{0.0175\ \text{C}}{\text{s}} \times \frac{1\ \text{F}}{96{,}485\ \text{C}} \times \frac{1\ \text{mol e}^-}{\text{F}} \times \frac{1\ \text{mol}}{2\ \text{mol e}^-} \times \frac{6.02 \times 10^{23}\ \text{ions}}{\text{mol}} = \frac{5.5 \times 10^{16}\ \text{ions}}{\text{s}}$$

22-14. (a)

$$E_{right} = 0.337 - \frac{0.0592}{2} \log\left(\frac{1}{0.250}\right) = 0.319 \text{ V}$$

$$E_{left} = 1.229 - \frac{0.0592}{4} \log\left(\frac{1}{1.00 \times \left(1.00 \times 10^{-3}\right)^4}\right) = 1.051 \text{ V}$$

$$E_{applied} = E_{right} - E_{left} = 0.319 - 1.051$$

$$= -0.732 \text{ V}$$

(c)

$$[H^+] = \text{antilog}\left(-3.70\right) = 1.995 \times 10^{-4}$$

$$E_{right} = 0.000 - \frac{0.0592}{2} \log\left(\frac{\frac{765}{760}}{\left(1.995 \times 10^{-4}\right)^2}\right) = -0.219 \text{ V}$$

$$E_{left} = 0.073 - 0.0592 \log\left(0.0964\right) = 0.133 \text{ V}$$

$$E_{applied} = E_{right} - E_{left} = -0.219 - 0.133$$

$$= -0.352 \text{ V}$$

22-15.

$$E_{right} = -0.763 - \frac{0.0592}{2} \log\left(\frac{1}{2.95 \times 10^{-3}}\right) = -0.838 \text{ V}$$

$$E_{left} = -0.277 - \frac{0.0592}{2} \log\left(\frac{1}{5.90 \times 10^{-3}}\right) = -0.343 \text{ V}$$

$$E_{cell} = -0.838 - \left(-0.343\right) - 0.065 \times 4.50$$

$$= -0.788 \text{ V}$$

22-17. (a)

$$E_{right} = 0.337 - \frac{0.0592}{2}\log\left(\frac{1}{0.250}\right) = 0.319 \text{ V}$$

$$E_{left} = 1.229 - \frac{0.0592}{4}\log\left(\frac{1}{\left(1.00\times10^{-4}\right)^4 \times \frac{730}{760}}\right) = 0.992 \text{ V}$$

$$E_{cell} = E_{right} - E_{left} = 0.319 - 0.992$$

$$= \quad -0.673 \text{ V}$$

(b) $IR = -0.15 \times 3.60 = -0.54 \text{ V}$

(c) Recall that the overpotential in an electrolytic cell requires the application of a larger or more negative potential. That is, 0.50 V must be subtracted from the cell potential.

$$E_{applied} = -0.673 - 0.54 - 0.50 = -1.71 \text{ V}$$

(d)

$$E_{right} = 0.337 - \frac{0.0592}{2}\log\left(\frac{1}{7.00\times10^{-6}}\right) = 0.184 \text{ V}$$

$$E_{applied} = 0.184 - 0.992 - 0.54 - 0.50$$

$$= -1.85 \text{ V}$$

22-19. Cd begins to form when

$$E = -0.403 - \frac{0.0592}{2}\log\left(\frac{1}{0.0650}\right) = -0.438 \text{ V}$$

(a) The Co^{2+} concentration when Cd first begins to deposit is:

$$-0.438 = -0.277 - \frac{0.0592}{2}\log\left(\frac{1}{[Co^{2+}]}\right)$$

$$\log\left([Co^{2+}]\right) = \frac{2\left(-0.438 + 0.277\right)}{0.0592} = -5.439$$

$$[Co^{2+}] = antilog\left(-5.439\right) = 3.6 \times 10^{-6} \text{ M}$$

(b) $E_{cathode} = -0.277 - \dfrac{0.0592}{2}\log\left(\dfrac{1}{1.00\times10^{-5}}\right) = -0.425\ \text{V}$

(c) Referring to Example 22-2, quantitative separation is assumed to occur when the $[Co^{2+}]$ falls to 10^{-4} of its original concentration or 2.0×10^{-5} M. Thus, if the cathode is maintained between -0.425 V and -0.438 V, the quantitative separation of Co^{2+} from Cd^{2+} is possible in theory.

22-21. (a) Bi deposits at a lower potential, that is

$[H^+] = \text{antilog}(-1.95) = 1.12\times10^{-2}\,\text{M}$

$E_{cathode} = 0.320 - \dfrac{0.0592}{3}\log\left(\dfrac{1}{0.250\left(1.12\times10^{-2}\right)^2}\right)$

$\qquad\qquad = 0.231\ \text{V}$

(b) Sn deposits when

$E_{cathode} = -0.136 - \dfrac{0.0592}{2}\log\left(\dfrac{1}{0.250}\right) = -0.154\ \text{V}$

$-0.154 = 0.320 - \dfrac{0.0592}{3}\log\left(\dfrac{1}{[BiO^+]\left(1.12\times10^{-2}\right)^2}\right)$

$\qquad = 0.320 + \dfrac{0.0592}{3}\log\left(1.12\times10^{-2}\right)^2 + \dfrac{0.0592}{3}\log\left([BiO^+]\right)$

$\log\left([BiO^+]\right) = \dfrac{3\left(-0.154-0.320+0.077\right)}{0.0592} = -20.12$

$[BiO^+] = \text{antilog}(-20.12) = 7.6\times10^{-21}\ \text{M}$

(c) When $[BiO^+] = 10^{-6}$

$E_{cathode} = 0.320 - \dfrac{0.0592}{3}\log\left(\dfrac{1}{1.00\times10^{-6}\left(1.12\times10^{-2}\right)^2}\right) = 0.124\ \text{V}$

Sn begins to form when $E_{cathode} = -0.154$ V (see part (b)) range vs. SCE = 0.124 − 0.244 to −0.154 − 0.244 or −0.12 to −0.398 V

= −0.120 to −0.398 V

22-22. Deposition of A is complete when

$$E_A = E_A^\circ - \frac{0.0592}{n_A} \log\left(\frac{1}{2.00 \times 10^{-5}}\right) = E_A^\circ - \frac{0.278}{n_A}$$

Deposition of B begins when

$$E_B = E_B^\circ - \frac{0.0592}{n_B} \log\left(\frac{1}{2.00 \times 10^{-1}}\right) = E_B^\circ - \frac{0.0414}{n_B}$$

Boundary condition is that $E_A = E_B$. Thus,

$$E_A^\circ - \frac{0.278}{n_A} = E_B^\circ - \frac{0.0414}{n_B} \quad \text{or}$$

$$E_A^\circ - E_B^\circ = \frac{0.278}{n_A} - \frac{0.0414}{n_B}$$

(a) $E_A^\circ - E_B^\circ = \dfrac{0.278}{1} - \dfrac{0.0414}{1} = 0.237$ V

(c) $E_A^\circ - E_B^\circ = \dfrac{0.278}{3} - \dfrac{0.0414}{1} = 0.0513$ V

(e) $E_A^\circ - E_B^\circ = \dfrac{0.278}{2} - \dfrac{0.0414}{2} = 0.118$ V

(g) $E_A^\circ - E_B^\circ = \dfrac{0.278}{1} - \dfrac{0.0414}{3} = 0.264$ V

(i) $E_A^\circ - E_B^\circ = \dfrac{0.278}{3} - \dfrac{0.0414}{3} = 0.0789$ V

22-23. (a)

$$0.250 \text{ g Co} \times \frac{1 \text{ mol Co}}{58.93 \text{ g}} \times \frac{2 \text{ mol e}^-}{\text{mol Co}} \times \frac{1 \text{ F}}{\text{mol e}^-} \times \frac{96,485 \text{ C}}{\text{F}} = 8.186 \times 10^2 \text{C}$$

$$8.186 \times 10^2 \text{ C} \times \frac{1 \text{ A} \times \text{s}}{\text{C}} \times \frac{1}{0.851 \text{ A}} \times \frac{1 \text{ min}}{60 \text{ s}} = 16.0 \text{ min}$$

(b) $3Co^{2+} + 4H_2O \rightleftharpoons Co_3O_4(s) + 8H^+ + 2e^-$ $\left(\dfrac{3}{2}\right)$ mol Co^{2+} = 1 mol e^-

$$0.250 \text{ g Co} \times \frac{1 \text{ mol Co}}{58.93 \text{ g}} \times \frac{2 \text{ mol } e^-}{3 \text{ mol Co}} \times \frac{1 \text{ F}}{\text{mol } e^-} \times \frac{96,485 \text{ C}}{\text{F}} = 2.727 \times 10^2 \text{ C}$$

$$2.727 \times 10^2 \text{ C} \times \frac{1 \text{ A} \times \text{s}}{\text{C}} \times \frac{1}{0.851 \text{ A}} \times \frac{1 \text{ min}}{60 \text{ s}} = 5.34 \text{ min}$$

22-25.

$$\left(5 \text{ min} \times \frac{60 \text{ s}}{\text{min}} + 24 \text{ s}\right) \times 0.300 \text{ A} \times \frac{1 \text{ C}}{\text{A} \times \text{s}} \times \frac{1 \text{ F}}{96,485 \text{ C}} \times \frac{1 \text{ eq HA}}{\text{F}} = 1.007 \times 10^{-3} \text{ eq HA}$$

$$\frac{0.1330 \text{ g HA}}{1.007 \times 10^{-3} \text{ eq HA}} = 132.0 \text{ g/eq}$$

22-27. 1 mol $CaCO_3$ = 1 mol $HgNH_3Y^{2-}$ = 2 mol e^-

$$\frac{\left(39.4 \times 10^{-3} \text{ A} \times 3.52 \text{ min} \times \dfrac{60 \text{ s}}{\text{min}} \times \dfrac{1 \text{ C}}{\text{A} \times \text{s}} \times \dfrac{1 \text{ mol } e^-}{96,485 \text{ C}} \times \dfrac{1 \text{ mol CaCO}_3}{2 \text{ mol } e^-} \times \dfrac{100.09 \text{ g CaCO}_3}{\text{mol}}\right)}{25.00 \text{ mL sample} \times \dfrac{1.00 \text{ g H}_2\text{O}}{\text{mL H}_2\text{O}}} \times 10^6 \text{ppm}$$

$$= 173 \text{ ppm CaCO}_3$$

22-29. 1 mol $C_6H_5NO_2$ = 4 mol e^-

$$\frac{\left(33.47 \text{ C} \times \dfrac{1 \text{ F}}{96,485 \text{ C}} \times \dfrac{1 \text{ mol } e^-}{\text{F}} \times \dfrac{1 \text{ mol C}_6\text{H}_5\text{NO}_2}{4 \text{ mol } e^-} \times \dfrac{123.11 \text{ g C}_6\text{H}_5\text{NO}_2}{\text{mol}}\right)}{300 \text{ mg sample} \times \dfrac{\text{g}}{1000 \text{ mg}}} \times 100\%$$

$$= 3.56\% \text{ C}_6\text{H}_5\text{NO}_2$$

171

23-34. 1 mol $C_6H_5NH_2$ = 3 mol Br_2 = 6 mol e^-

$$\left(\begin{array}{c} (3.76 - 0.27)\,min \;\times\; \dfrac{60\,s}{min} \;\times\; \dfrac{1.51\times10^{-3}\,C}{s} \;\times\; \dfrac{1\,F}{96,485\,C} \times \\[2mm] \dfrac{1\,mol\,e^-}{F} \;\times\; \dfrac{1\,mol\,C_6H_5NH_2}{6\,mol\,e^-} \end{array} \right) = 5.462 \times 10^{-7}\,mol\,C_6H_5NH_2$$

$$5.462 \times 10^{-7}\,mol\,C_6H_5NH_2 \times \dfrac{93.128\,g\,C_6H_5NH_2}{mol} \times \dfrac{10^6\,\mu g}{g}$$

$$= 50.9\,\mu g\,C_6H_5NH_2$$

23-35. 1 mol Sn^{4+} = 2 mol $e^- \rightarrow$ 1 mol Sn^{2+} = 2 mol $C_6H_4O_2$

$$\left(\begin{array}{c} (8.34 - 0.691)\,min \times \dfrac{60\,s}{min} \;\times\; \dfrac{1.062\times10^{-3}\,C}{s} \times \dfrac{1\,F}{96,485\,C} \\[2mm] \times\dfrac{1\,mol\,e^-}{F} \times \dfrac{1\,mol\,C_6H_5NH_2}{2\,mol\,e^-} \end{array} \right) = 2.526 \times 10^{-6}\,mol\,C_6H_4O_2$$

$$2.526 \times 10^{-6}\,mol\,C_6H_4O_2 \times \dfrac{108.10\,g\,C_6H_4O_2}{mol}$$

$$= 2.73 \times 10^{-4}\,g\,C_6H_4O_2$$

Chapter 23

23-1. **(a)** *Voltammetry* is an analytical technique that is based on measuring the current that

develops at a small electrode as the applied potential is varied. *Amperometry* is a

technique in which the limiting current is measured at a constant potential.

(c) *Differential pulse* and *square wave voltammetry* differ in the type of pulse sequence

used as shown in Figure 23-1b and 23-1c.

(e) In voltammetry, a *limiting current* is a current that is independent of applied potential.

Its magnitude is limited by the rate at which a reactant is brought to the surface of the

electrode by migration, convection, and/or diffusion. A *diffusion current* is a limiting

current when analyte transport is solely by diffusion.

(g) The *half-wave potential* is closely related to the *standard potential* for a reversible

reaction. That is,

$$E_{1/2} = E_A^\circ - \frac{0.0592}{n} \log\left(\frac{k_A}{k_B}\right)$$

where k_A and k_B are constants that are proportional to the diffusion coefficients of the

analyte and product. When these are approximately the same, the half-wave potential

and the standard potential are essentially equal.

23-3. A high supporting electrolyte concentration is used in most electroanalytical procedures

to minimize the contribution of migration to concentration polarization. The supporting

electrolyte also reduces the cell resistance, which decreases the *IR* drop.

23-5. Most organic electrode processes consume or produce hydrogen ions. Unless buffered

solutions are used, marked pH changes can occur at the electrode surface as the reaction

proceeds.

23-7. The purpose of the electrodeposition step in stripping analysis is to preconcentrate the analyte on the surface of the working electrode and to separate it from many interfering species.

23-9. A plot of E_{appl} versus $\log \dfrac{i}{i_l - i}$ should yield a straight line having a slope of $\dfrac{-0.0592}{n}$.

Thus, n is readily obtained from the slope.

23-12. Initally there are 60 mL $\times$ 0.08 mmol/mL = 4.8 mmol Cu^{2+} present.

Applying a current of 6.0 µA for 45 minutes represents a charge of

6.0×10^{-6} C/s $\times$ 45 min $\times$ 60 s/min = 0.0162 C

The number of moles of Cu^{2+} reduced by that amount of charge is:

$n_{Cu2+} = Q/nF = 0.0162$ C/$(2 \times 96485$ C/mol$) = 8.4 \times 10^{-8}$ mol or 8.4×10^{-5} mmol

The percentage removed is thus $(8.4 \times 10^{-5}$ mmol$/ 4.8$ mmol$) \times 100\% = 1.7 \times 10^{-3}\%$

23-13. $i_1 = kc_u$ where $i_1 = 1.86$ µA and c_u is the concentration of the unknown.

$$i_2 = \frac{k\left(25.00c_u + 5.00 \times 2.12 \times 10^{-3}\right)}{25.00 + 5.00} = 5.27 \text{ µA}$$

From above, $k = i_1/c_u$. Substituting this into the second equation and solving for c_u gives

$c_u = 1.77 \times 10^{-4}$ M

Chapter 24

24-1. The yellow color comes about because the solution absorbs blue light in the wavelength

region 435-480 nm and transmits its complementary color (yellow). The purple color

comes about because green radiation (500-560 nm) is absorbed and its complementary

color (purple) is transmitted.

24-2. (a) Absorbance A is the negative logarithm of transmittance T ($A = -\log T$).

24-3. Deviations from linearity can occur because of polychromatic radiation, unknown

chemical changes such as association or dissociation reactions, stray light, and molecular

or ionic interactions at high concentration.

24-4. $v = c/\lambda = 3.00 \times 10^{10}$ cm s^{-1}/λ(cm) = $(3.00 \times 10^{10}/\lambda)$ s^{-1} = $(3.00 \times 10^{10}/\lambda)$ Hz

(a) $v = 3.00 \times 10^{10}$ cm s^{-1}/(2.65 Å $\times 10^{-8}$ cm/Å) = 1.13×10^{18} Hz

(c) $v = 3.00 \times 10^{10}$ cm s^{-1}/(694.3 nm $\times 10^{-7}$ cm/nm) = 4.32×10^{14} Hz

(e) $v = 3.00 \times 10^{10}$ cm s^{-1} / (19.6 μm $\times 10^{-4}$ cm/μm) = 1.53×10^{13} Hz

24-5. $\lambda = c/v = 3.00 \times 10^{10}$ cm s^{-1}/v (s^{-1}) = $(3.00 \times 10^{10}/v)$ cm

(a) $\lambda = 3.00 \times 10^{10}$ cm s^{-1}/(118.6 MHz $\times 10^6$ Hz/MHz) = 253.0 cm

(c) $\lambda = 3.00 \times 10^{10}$ cm s^{-1}/(105 MHz $\times 10^6$ Hz/MHz) = 286 cm

24-6. (a) $\bar{V} = 1/(3$ μm $\times 10^{-4}$ cm/μm) = 3.33×10^3 cm^{-1} to

$$1/(15 \times 10^{-4} \text{ cm}) = 6.67 \times 10^2 \text{ cm}^{-1}$$

(b) $v = 3.00 \times 10^{10}$ cm s^{-1} $\times 3.333 \times 10^3$ cm^{-1} = 1.00×10^{14} Hz to

$$3.00 \times 10^{10} \times 6.67 \times 10^2 = 2.00 \times 10^{13} \text{ Hz}$$

24-7. $\lambda = c/v = (3.00 \times 10^{10}$ cm s$^{-1})$ / $(220 \times 10^6$ s$^{-1})$ = 136 cm or 1.36 m

$E = hv = 6.63 \times 10^{-34}$ J s $\times 220 \times 10^6$ s^{-1} = 1.46×10^{-25} J

24-8.   (a) $\lambda = 589$ nm/1.35 = 436 nm

24-9.   (a) ppm^{-1} cm^{-1}

(c) %$^{-1}$ cm^{-1}

24-10. (a) %$T = 100 \times$ antilog(−0.0356) = 92.1%

Proceeding similarly, we obtain

(c) %$T = 41.8$; (e) %$T = 32.7$

24-11. (a) $A = -\log T = -\log(27.2\%/100\%) = 0.565$

Proceeding similarly,

(c) $A = 0.514$; (e) $A = 1.032$

24-12. (a) %$T =$ antilog(−0.172) × 100% = 67.3%

$$c = A/\varepsilon b = (0.172)/(4.23 \times 10^3 \times 1.00) = 4.07 \times 10^{-5} \text{ M}$$

$$c = 4.07 \times 10^{-5}\frac{\text{mol}}{\text{L}} \times \frac{200 \text{ g}}{\text{mol}} \times \frac{1 \text{ L}}{1000 \text{ g}} \times 10^6 \text{ ppm} = 8.13 \text{ ppm}$$

$$a = A/bc = 0.172/(1.00 \times 8.13) = 0.0211 \text{ cm}^{-1} \text{ ppm}^{-1}$$

Using similar conversions and calculations, we can evaluate the missing quantities

	A	%T	ε L mol^{-1} cm^{-1}	a cm^{-1} ppm^{-1}	b cm	c M	ppm
*(a)	0.172	67.3	4.23×10^3	0.0211	1.00	4.07×10^{-5}	8.13
*(c)	0.520	30.2	7.95×10^3	0.0397	1.00	6.54×10^{-5}	13.1
*(e)	0.638	23.0	3.73×10^3	0.0187	0.100	1.71×10^{-3}	342
*(g)	0.798	15.9	3.17×10^3	0.0158	1.50	1.68×10^{-4}	33.6
*(i)	1.28	5.23	9.78×10^3	0.0489	5.00	2.62×10^{-5}	5.24

24-13. (a) $A = 7.00 \times 10^3$ L mol^{-1} cm$^{-1} \times 1.00$ cm $\times 3.40 \times 10^{-5}$ mol L$^{-1} = 0.238$

(b) $A = 7.00 \times 10^3 \times 1.00 \times 2 \times 3.40 \times 10^{-5} = 0.476$

(c) For part (a), $T = $ antilog$(-0.238) = 0.578$

For part (b), $T = $ antilog$(-0.476) = 0.334$

(d) $A = -\log(T) = -\log(0.578/2) = 0.539$

24-14. (a) $A = 9.32 \times 10^3$ L mol^{-1} cm$^{-1} \times 1.00$ cm $\times 5.67 \times 10^{-5}$ mol L$^{-1} = 0.528$

(b) $\%T = 100 \times $ anitlog$(-0.528) = 29.6\%$

(c) $c = A/\varepsilon b = 0.528/(9.32 \times 10^3$ L mol^{-1} cm$^{-1} \times 2.50$ cm$) = 2.27 \times 10^{-5}$ M

24-15. $2.10 = -\log(P/P_0)$ $\qquad$ $P/P_0 = 0.0079433$

$$P = 0.007943\, P_0$$

$$P_s/P_0 = 0.0075$$

$$P_s = 0.0075\, P_0$$

$$A' = \left(\frac{P_0 + P_s}{P + P_s}\right) = \log\left(\frac{P_0 + 0.0075 P_0}{0.007943 P_0 + 0.0075 P_0}\right) = \log\left(\frac{1.0075 P_0}{0.015443 P_0}\right) = \log(65.2139)$$

$$= 1.81$$

Error $= [(1.81 - 2.10)/2.10] \times \%100 = -13.6\%$

Chapter 25

25-1. (a) *Phototubes* consist of a single photoemissive surface (cathode) and an anode in an

evacuated envelope. They exhibit low dark current, but have no inherent amplification.

Solid-state photodiodes are semiconductor *pn*-junction devices that respond to incident

light by forming electron-hole pairs. They are more sensitive than phototubes but less

sensitive than photomultiplier tubes.

(c) *Filters* isolate a single band of wavelengths. They provide low resolution wavelength

selection suitable for quantitative work. *Monochromators* produce high resolution for

qualitative and quantitative work. With monochromators, the wavelength can be varied

continuously, whereas this is not possible with filters.

25-2. Quantitative analyses can tolerate rather wide slits since measurements are usually

carried out at a wavelength maximum where the slope of the spectrum $dA/d\lambda$ is relatively

constant. On the other hand, qualitative analyses require narrow slits so that any fine

structure in the spectrum will be resolved. This can allow differentiation of one

compound from another.

25-3. *Tungsten/halogen lamps* contain a small amount of iodine in the evacuated quartz

envelope that contains the tungsten filament. The iodine prolongs the life of the lamp and

permits it to operate at a higher temperature. The iodine combines with gaseous tungsten

that sublimes from the filament and causes the metal to be redeposited, thus adding to the

life of the lamp.

25-4. **(a)** *Spectrophotometers* have monochromators for multiple wavelength operation and for

procuring spectra while *photometers* utilize filters for fixed wavelength operation. While

offering the advantage of multiple wavelength operation, spectrophotometers are

substantially more complex and more expensive than photometers.

(c) Both a *monochromator* and a *polychromator* use a diffraction grating to disperse the

spectrum, but a monochromator contains only one exit slit and detector while a

polychromator contains multiple exit slits and detectors. A monochromator can be used

to monitor one wavelength at a time while a polychromator can monitor several discrete

wavelengths simultaneously.

25-5. **(a)** $\lambda_{max} = 2.90 \times 10^3 / T = 2.90 \times 10^3 / 4000 = 0.73 \ \mu m$

(c) $\lambda_{max} = 2.90 \times 10^3 / 2000 = 1.45 \ \mu m$

25-6. **(a)** $\lambda_{max} = 2.90 \times 10^3 / 2870 = 1.01 \ \mu m \ (1010 \ nm)$

$\lambda_{max} = 2.90 \times 10^3 / 3000 = 0.967 \ \mu m \ (967 \ nm)$

(b) $E_t = 5.69 \times 10^{-8} (2870)^4 \times (1 \ m / 100 \ cm)^2 = 386 \ W/cm^2$

$E_t = 5.69 \times 10^{-8} (3000)^4 \times (1 \ m / 100 \ cm)^2 = 461 \ W/cm^2$

25-7. **(a)** The 0% transmittance is measured with no light reaching the detector and is a

measure of the dark current.

(b) The 100% transmittance adjustment is made with a blank in the light path and

measures the unattenuated source. It compensates for any absorption or reflection losses

in the cell and optics.

25-8. Fourier transform IR spectrometers have the advantages over dispersive instruments of higher speed and sensitivity, better light-gathering power, more accurate and precise wavelength settings, simpler mechanical design, and elimination of stray light and IR emission.

25-9. **(a)** $\%T = (149 / 625) \times 100\% = 23.84\%$

$$A = -\log(23.84\% / 100\%) = 0.623$$

(c) Since A is proportional to light path, at twice the light path $A = 2 \times 0.623 = 1.246$

$$T = \text{antilog}(-A) = \text{antilog}(-1.246) = 0.057; \ \%T = 5.7$$

25-10. **(b)** $A = -\log(30.96\%/100\%) = 0.509$

(d) $A = 2 \times 0.509 = 1.018$

$$T = \text{antilog}(-A) = \text{antilog}(-1.018) = 0.096$$

25-11. A *photon detector* produces a current or voltage as a result of the emission of electrons from a photosensitive surface when struck by photons. A *thermal detector* consists of a darkened surface to absorb infrared radiation and produce a temperature increase. A *thermal transducer* produces an electrical signal whose magnitude is related to the temperature and thus the intensity of the infrared radiation.

25-12. Basically, an *absorption photometer* and a *fluorescence photometer* consist of the same components. The basic difference is in the location of the detector. The detector in a fluorometer is positioned at an angle of 90° to the direction of the beam from the source so that emission is detected rather than transmisson. In addition, a filter is often positioned in front of the detector to remove radiation from the excitation beam that may result from scattering or other nonfluorescence processes. In a transmission photometer, the detector is positioned in a line with the source, the filter, and the detector.

25-13. **(a)** *Transducer* indicates the type of detector that converts quantities, such as light

intensity, pH, mass, and temperature, into electrical signals that can be subsequently

amplified, manipulated, and finally converted into numbers proportional to the magnitude

of the original quantity.

(c) A semiconductor containing unbonded electrons (e.g. produced by doping silicon

with a Group V element) is termed an *n*-type semiconductor.

(e) A *depletion layer* results when a reverse bias is applied to a *pn*-junction type device.

Majority carriers are drawn away from the junction leaving a nonconductive depletion

layer.

Chapter 26

26-1. **(a)** *Spectrophotometers* use a grating or a prism to provide narrow bands of radiation

while *photometers* use filters for this purpose. The advantages of spectrophotometers are

greater versatility and the ability to obtain entire spectra. The advantages of photometers

are simplicity, ruggedness, higher light throughput and low cost.

(c) *Diode-array spectrophotometers* detect the entire spectral range essentially

simultaneously and can produce a spectrum in less than a second. *Conventional*

spectrophotometers require several minutes to scan the spectrum. Accordingly, diode-

array instruments can be used to monitor processes that occur on fast time scales. Their

resolution is usually lower than that of a conventional spectrophotometer.

26-2. Electrolyte concentration, pH, temperature, nature of solvent, and interfering substances.

26-3. $A = \varepsilon\, b\, c$

$c_{min} = A/\varepsilon b = 0.10/(9.32 \times 10^3 \times 1.00) = 1.1 \times 10^{-5} \text{ M}$

$c_{max} = A/\varepsilon b = 0.90/(9.32 \times 10^3 \times 1.00) = 9.7 \times 10^{-5} \text{ M}$

26-4. $\log \varepsilon = 2.75$ $\qquad\qquad \varepsilon = 5.6 \times 10^2$

$c_{min} = A/\varepsilon b = 0.100/(5.6 \times 10^2 \times 1.50) = 1.2 \times 10^{-4} \text{ M}$

$c_{max} = A/\varepsilon b = 2.000/(5.6 \times 10^2 \times 1.50) = 2.4 \times 10^{-3} \text{ M}$

26-5. **(a)** $T = 169 \text{ mV}/690 \text{ mV} = 0.245$

$A = -\log(0.245) = 0.611$

(c) Since A is proportional to light path, at twice the light path $A = 2 \times 0.611 = 1.222$

$T = \text{antilog}(-A) = \text{antilog}(-1.222) = 0.060$

26-6. **(b)** $A = -\log(31.4\% / 100\%) = 0.503$

(d) $A = 2 \times 0.503 = 1.006$

$T = \text{antilog}(-A) = \text{antilog}(-1.006) = 0.099$

26-7.

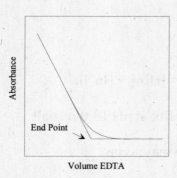

The absorbance should decrease approximately linearly with titrant volume until the end point.

After the end point the absorbance becomes independent of titrant volume.

26-8. Applying the equation we developed in Solution 26-15 we write

$$c_x = \frac{0.231 \times 2.75 \times 5.00}{50.0(0.549 - 0.231)} = 0.200 \text{ ppm Fe}$$

26-9.

$$A_{510} = 0.446 = 36400 \times 1.00 \times c_{Co} + 5520 \times 1.00 \times c_{Ni}$$

$$A_{656} = 0.326 = 1240 \times 1.00 \times c_{Co} + 17500 \times 1.00 \times c_{Ni}$$

$$c_{Co} = 9.530 \times 10^{-6} \text{ M} \qquad\qquad c_{Ni} = 1.795 \times 10^{-5} \text{ M}$$

$$c_{Co} = \frac{50.0 \text{ mL} \times 9.530 \times 10^{-6} \dfrac{\text{mmol}}{\text{mL}} \times \dfrac{50.0 \text{ mL}}{25.0 \text{ mL}} \times \dfrac{0.05893 \text{ g Co}}{\text{mmol}}}{0.425 \text{ g}} \times 10^6 \text{ ppm} = 132 \text{ ppm}$$

$$c_{Ni} = \frac{50.0 \text{ mL} \times 1.795 \times 10^{-5} \dfrac{\text{mmol}}{\text{mL}} \times \dfrac{50.0 \text{ mL}}{25.0 \text{ mL}} \times \dfrac{0.05869 \text{ g Ni}}{\text{mmol}}}{0.425 \text{ g}} \times 10^6 \text{ ppm} = 248 \text{ ppm}$$

26-10. $\alpha_0 = \dfrac{[H_3O^+]}{[H_3O^+] + K_{HIn}}$ $\alpha_1 = 1 - \alpha_0$

$A_{450} = \varepsilon_{HIn} \times 1.00 \times [HIn] + \varepsilon_{In} \times 1.00 \times [In^-]$

$\qquad = \varepsilon_{HIn}\alpha_0 c_{In} + \varepsilon_{In}\alpha_1 c_{In}$

$\qquad = (\varepsilon_{HIn}\alpha_0 + \varepsilon_{In}\alpha_1)c_{In}$

where c_{In} is the analytical concentration of the indicator ($c_{In} = [HIn] + [In^-]$).

We may assume at pH 1.00 all of the indicator is present as HIn; at pH 13.0 it is all

present as In$^-$. Therefore, from the data in Problem 26-19 we may write

$\varepsilon_{HIn} = \dfrac{A_{450}}{bc_{HIn}} = \dfrac{0.658}{1.00 \times 8.00 \times 10^{-5}} = 8.22 \times 10^3 \text{ L mol}^{-1}\text{ cm}^{-1}$

$\varepsilon_{In} = \dfrac{A_{450}}{bc_{In}} = \dfrac{0.076}{1.00 \times 8.00 \times 10^{-5}} = 9.5 \times 10^2 \text{ L mol}^{-1}\text{ cm}^{-1}$

(a)   At pH = 4.92, $[H_3O^+] = 1.20\times10^{-5}$ M

$\alpha_0 = \dfrac{1.20 \times 10^{-5}}{1.20 \times 10^{-5} + 4.80 \times 10^{-6}} = 0.714$

$\alpha_1 = 1.000 - 0.714 = 0.286$

$A_{450} = (8.22 \times 10^3 \times 0.714 + 9.5 \times 10^2 \times 0.286) \times 8.00 \times 10^{-5} = 0.492$

	pH	[H$_3$O$^+$]	α_0	α_1	A_{450}
(a)	4.92	1.20×10^{-5}	0.714	0.286	0.492
(c)	5.93	1.18×10^{-6}	0.197	0.803	0.190

26-11. The approach is identical to that of Solution 26-20. At 595 nm and

at pH = 1.00, $\varepsilon_{HIn} = \dfrac{A_{595}}{bc_{HIn}} = \dfrac{0.032}{1.00 \times 8.00 \times 10^{-5}} = 4.0 \times 10^2 \text{ L mol}^{-1}\text{ cm}^{-1}$

at pH = 13.00, $\varepsilon_{In} = \dfrac{A_{595}}{bc_{In}} = \dfrac{0.361}{1.00 \times 8.00 \times 10^{-5}} = 4.51 \times 10^3 \text{ L mol}^{-1}\text{ cm}^{-1}$

(a) At pH = 5.30 and with 1.00-cm cells, $[H_3O^+] = 5.01 \times 10^{-6}$ M and

$$\alpha_0 = \frac{[H_3O^+]}{[H_3O^+] + K_{HIn}} = \frac{5.01 \times 10^{-6}}{5.01 \times 10^{-6} + 4.80 \times 10^{-6}} = 0.511$$

$$\alpha_1 = 1 - \alpha_0 = 0.489$$

$$A_{595} = (\varepsilon_{HIn}\alpha_0 + \varepsilon_{In}\alpha_1)c_{In}$$

$$A_{595} = (4.0 \times 10^2 \times 0.511 + 4.51 \times 10^3 \times 0.489) \times 1.25 \times 10^{-4} = 0.301$$

Similarly for parts (b) and (c)

	pH	$[H_3O^+]$	α_0	α_1	A_{595}
(a)	5.30	5.01×10^{-6}	0.511	0.489	0.301
(b)	5.70	2.00×10^{-6}	0.294	0.706	0.413
(c)	6.10	7.94×10^{-7}	0.142	0.858	0.491

26-12. In these solutions the concentrations of the two absorbers HIn and In$^-$ must be determined

by the analysis of mixtures, so

$$A_{450} = \varepsilon'_{HIn}b[HIn] + \varepsilon'_{In}b[In^-]$$

$$A_{595} = \varepsilon''_{HIn}b[HIn] + \varepsilon''_{In}b[In^-]$$

From the solutions to 26-20 and 26-21

$$\varepsilon'_{HIn} = 8.22 \times 10^3 \quad \varepsilon'_{In} = 9.5 \times 10^2 \quad \varepsilon''_{HIn} = 4.0 \times 10^2 \quad \varepsilon''_{In} = 4.51 \times 10^3$$

Thus, $A_{450} = 0.344 = (8.22 \times 10^3)[HIn] + (9.5 \times 10^2)[In^-]$

$$A_{595} = 0.310 = (4.0 \times 10^2)[HIn] + (4.51 \times 10^3)[In^-]$$

Solving these equations gives

$$[HIn] = 3.42 \times 10^{-5} \text{ M} \qquad \text{and} \qquad [In^-] = 6.57 \times 10^{-5} \text{ M}$$

$$K_{HIn} = \frac{[H_3O^+][In^-]}{[HIn]}$$

$$[H_3O^+] = K_{HIn} \frac{[HIn]}{[In^-]} = \frac{(4.80 \times 10^{-6})(3.42 \times 10^{-5})}{6.57 \times 10^{-5}} = 2.50 \times 10^{-6} \text{ M}$$

$$pH = -\log[H_3O^+] = -\log(2.50 \times 10^{-6}) = 5.60$$

The results for all solutions are shown in the table that follows.

Solution	[HIn]	[In$^-$]	pH
A	3.42×10^{-5}	6.57×10^{-5}	5.60
C	7.70×10^{-5}	2.33×10^{-5}	4.80

26-13.　　　$A_{440} = \varepsilon_P' b c_P + \varepsilon_Q' b c_Q$　　　　$b = 1.00$ cm

$$A_{620} = \varepsilon_P'' b c_P + \varepsilon_Q'' b c_Q$$

$$c_P = \frac{A_{440} - \varepsilon_Q' c_Q}{\varepsilon_P'}$$

Substituting for c_P in the second equation gives

$$A_{620} = \varepsilon_P'' \left[\frac{A_{440} - \varepsilon_Q' c_Q}{\varepsilon_P'} \right] + \varepsilon_Q'' c_Q$$

We then solve for c_Q and c_P as in the spreadsheet

▲	A	B	C	D	E
1		ε(P), M^{-1}cm-1	ε(Q), M^{-1}cm-1		
2	440	1.123E+03	2.527E+03		
3	620	3.813E+03	2.321E+02		
4					
5		A_{440}	A_{620}	[P], M	[Q], M
6	(a)	0.357	0.803	2.076E-04	4.901E-05
7	(b)	0.830	0.448	1.002E-04	2.839E-04
8	(c)	0.248	0.333	8.362E-05	6.098E-05
9	(d)	0.910	0.338	6.858E-05	3.296E-04
10	(e)	0.480	0.825	2.105E-04	9.640E-05
11	(f)	0.194	0.315	8.011E-05	4.117E-05
12	**Documentation**				
13	B2:C3 From Problem 26-24				
14	D6=(B6-$C$2*C6/$C$3)/($B$2-$C$2*$B$3/$C$3)				
15	E6=(B6-$B$2*D6)/$C$2				

26-14.

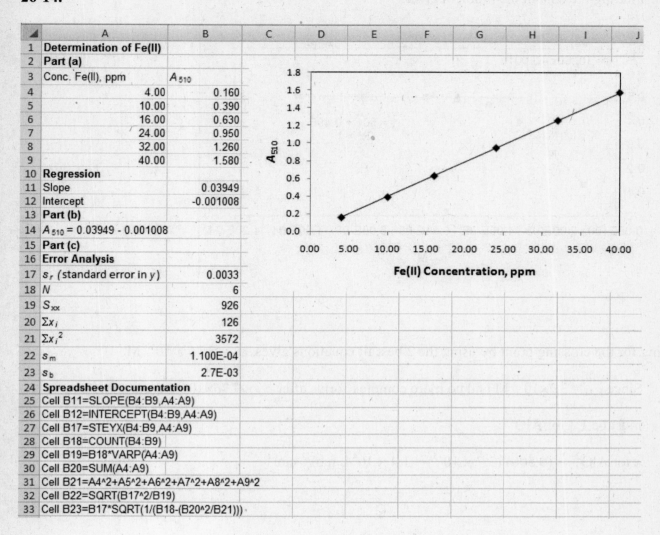

	A	B	C	D	E	F	G	H	I	J
1	**Determination of Fe(II)**									
2	**Part (a)**									
3	Conc. Fe(II), ppm	A_{510}								
4	4.00	0.160								
5	10.00	0.390								
6	16.00	0.630								
7	24.00	0.950								
8	32.00	1.260								
9	40.00	1.580								
10	**Regression**									
11	Slope	0.03949								
12	Intercept	-0.001008								
13	**Part (b)**									
14	A_{510} = 0.03949 - 0.001008									
15	**Part (c)**									
16	**Error Analysis**									
17	s_r (standard error in y)	0.0033								
18	N	6								
19	S_{xx}	926								
20	Σx_i	126								
21	Σx_i^2	3572								
22	s_m	1.100E-04								
23	s_b	2.7E-03								
24	**Spreadsheet Documentation**									
25	Cell B11=SLOPE(B4:B9,A4:A9)									
26	Cell B12=INTERCEPT(B4:B9,A4:A9)									
27	Cell B17=STEYX(B4:B9,A4:A9)									
28	Cell B18=COUNT(B4:B9)									
29	Cell B19=B18*VARP(A4:A9)									
30	Cell B20=SUM(A4:A9)									
31	Cell B21=A4^2+A5^2+A6^2+A7^2+A8^2+A9^2									
32	Cell B22=SQRT(B17^2/B19)									
33	Cell B23=B17*SQRT(1/(B18-(B20^2/B21)))									

26-15. Plotting the data in the problem gives

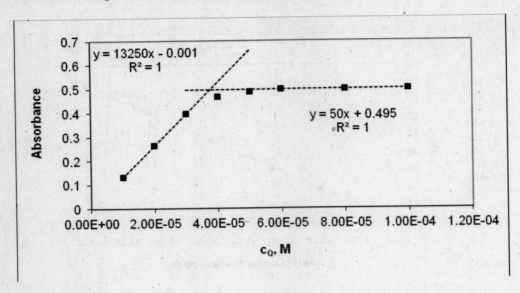

Solving for the crossing point by using the 2 best fit equations gives, $c_Q = 3.76 \times 10^{-5}$ M.

(a) Since $c_{Al} = 3.7 \times 10^{-5}$ M and no more complex forms after $c_Q = 3.76 \times 10^{-5}$ M, the complex

must be 1:1, or AlQ^{2+}.

(b) ε for $AlQ^{2+} = (0.500)/(3.7 \times 10^{-5}) = 1.4 \times 10^4$ L mol^{-1} cm^{-1}

26-16.

	A	B	C	D	E	F	G	H	I	J	K	L	M
1	Solution	V_M	V_L	$V_M/(V_M+V_L)$	A_{390}								
2	0	10.00	0.00	1.00	0.000								
3	1	9.00	1.00	0.90	0.174								
4	2	8.00	2.00	0.80	0.353								
5	3	7.00	3.00	0.70	0.530								
6	4	6.00	4.00	0.60	0.672								
7	5	5.00	5.00	0.50	0.723								
8	6	4.00	6.00	0.40	0.673								
9	7	3.00	7.00	0.30	0.537								
10	8	2.00	8.00	0.20	0.358								
11	9	1.00	9.00	0.10	0.180								
12	10	0.00	10.00	0.00	0.000								
13													
14	Slope 1	1.789											
15	Intercept 1	0.0004											
16	Slope 2	-1.769											
17	Intercept 2	1.7679											
18	1.789x+1.769x = 1.769-0.0004												
19	x	0.497											
20	c_{Cd2+}	1.25E-04 M											
21	c_R	1.25E-04 M											
22	b	1.00 cm											
23													
24		ε values											
25	Soln1	13920.00											
26	Soln2	14120.00											
27	Soln3	14133.33											
28	Soln7	14320.00											
29	Soln8	14320.00											
30	Soln9	14400.00											
31													
32	Average	14202											
33	SD	162											

Chart: A_{390} vs $V_M/(V_M+V_L)$, with lines $y = 1.789x + 0.0004$ and $y = -1.769x + 1.7679$.

(a) The two lines intercept at $V_M / (V_M + V_L) = 0.5$ (Cell B19). The Cd^{2+} to R ratio is 1:1.

(b) The molar absorptivities can be obtained from solutions 1-3 where the reagent is limiting and solutions 7-9 where the metal is limiting. Rounding the results in Cells B32 and B33 the average $\varepsilon = 1400 \pm 200$ L mol^{-1} cm^{-1}

(c) The absorbance at the volume ratio where the lines intersect is $A = 0.723$. Thus,

$$[CdR] = (0.723)/(14202) = 5.09 \times 10^{-5} \text{ M}$$

$$[Cd^{2+}] = [(5.00 \text{ mL})(1.25 \times 10^{-4} \text{ mmol/mL}) - (10.00 \text{ mL})(5.09 \times 10^{-5} \text{ mmol/mL})]/(10.00 \text{ mL})$$

$$= 1.16 \times 10^{-5} \text{ M}$$

189

$$[R] = [Cd^{2+}] = 1.16 \times 10^{-5} \text{ M}$$

$$K_f = \frac{[CdR]}{[Cd^{2+}][R]} = \frac{(5.09 \times 10^{-5})}{(1.16 \times 10^{-5})^2} = 3.78 \times 10^5$$

26-17. From Figure 26F-2, the frequencies of the band maxima are estimated to be:

(1) 740 cm^{-1} C-Cl stretch

(2) 1270 cm^{-1} CH$_2$ wagging

(3) 2900 cm^{-1} Aliphatic C-H stretch.

Chapter 27

27-1. **(a)** *Fluorescence* is a photoluminescence process in which atoms or molecules are

excited by absorption of electromagnetic radiation and then relax to the ground state,

giving up their excess energy as photons. The transition is from the lowest lying excited

singlet state to the ground singlet state.

(c) *Internal conversion* is the nonradiative relaxation of a molecule from a low energy

vibrational level of an excited electronic state to a high energy vibrational level of a

lower electronic state.

(e) The *Stokes shift* is the difference in wavelength between the radiation used to excite

fluorescence and the wavelength of the emitted radiation.

(g) An *inner filter effect* is a result of excessive absorption of the incident beam (primary

absorption) or absorption of the emitted beam (secondary absorption).

27-2. **(a)** Fluorescein because of its greater structural rigidity due to the bridging –O– groups.

27-3. Organic compounds containing aromatic rings often exhibit fluorescence. Rigid

molecules or multiple ring systems tend to have large quantum yields of fluorescence

while flexible molecules generally have lower quantum yields.

27-4. See Figure 27-8. A filter fluorometer usually consists of a light source, a filter for

selecting the excitation wavelength, a sample container, an emission filter and a

transducer/readout device. A spectrofluorometer has two monochromators that are the

wavelength selectors.

27-5. Fluorometers are more sensitive because filters allow more excitation radiation to reach

the sample and more emitted radiation to reach the transducer. Thus, a fluorometer can

provide lower limits of detection than a spectrofluorometer. In addition, fluorometers are substantially less expensive and more rugged than spectrofluorometer, making them

particularly well suited for routine quantitation and remote analysis applications.

27-6.

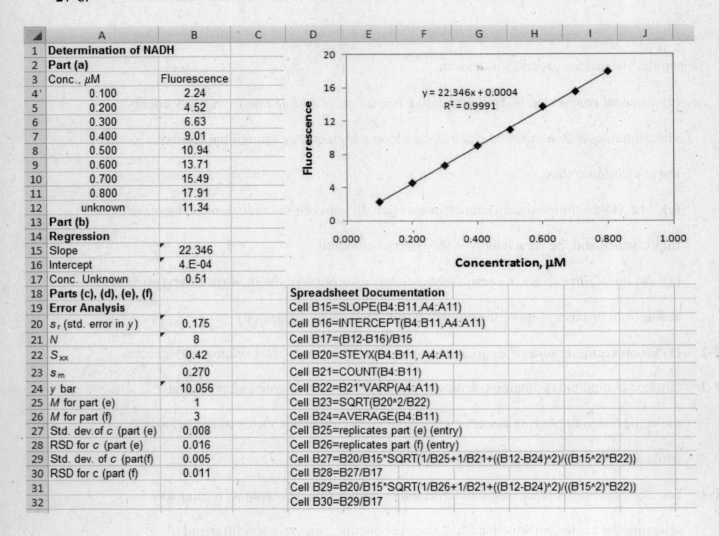

	A	B	C	D	E	F	G	H	I	J
1	**Determination of NADH**									
2	**Part (a)**									
3	Conc., μM	Fluorescence								
4	0.100	2.24								
5	0.200	4.52								
6	0.300	6.63								
7	0.400	9.01								
8	0.500	10.94								
9	0.600	13.71								
10	0.700	15.49								
11	0.800	17.91								
12	unknown	11.34								
13	**Part (b)**									
14	**Regression**									
15	Slope	22.346								
16	Intercept	4.E-04								
17	Conc. Unknown	0.51								
18	**Parts (c), (d), (e), (f)**			**Spreadsheet Documentation**						
19	**Error Analysis**			Cell B15=SLOPE(B4:B11,A4:A11)						
20	s_r (std. error in y)	0.175		Cell B16=INTERCEPT(B4:B11,A4:A11)						
21	N	8		Cell B17=(B12-B16)/B15						
22	S_{xx}	0.42		Cell B20=STEYX(B4:B11, A4:A11)						
23	s_m	0.270		Cell B21=COUNT(B4:B11)						
24	y bar	10.056		Cell B22=B21*VARP(A4:A11)						
25	M for part (e)	1		Cell B23=SQRT(B20^2/B22)						
26	M for part (f)	3		Cell B24=AVERAGE(B4:B11)						
27	Std. dev.of c (part (e)	0.008		Cell B25=replicates part (e) (entry)						
28	RSD for c (part (e)	0.016		Cell B26=replicates part (f) (entry)						
29	Std. dev. of c (part(f)	0.005		Cell B27=B20/B15*SQRT(1/B25+1/B21+((B12-B24)^2)/((B15^2)*B22))						
30	RSD for c (part (f)	0.011		Cell B28=B27/B17						
31				Cell B29=B20/B15*SQRT(1/B26+1/B21+((B12-B24)^2)/((B15^2)*B22))						
32				Cell B30=B29/B17						

27-7.　$c_Q = 100 \text{ ppm} \times 288/180 = 160 \text{ ppm}$

$$160 \text{ ppm} \times \frac{100 \text{ mL}}{15 \text{ mL}} \times \frac{1 \text{ mg quinine}}{1 \times 10^3 \text{ g solution}} \times \frac{1 \text{ g solution}}{1 \text{ mL}} \times 500 \text{ mL} = 533 \text{ mg quinine}$$

Chapter 28

28-1.   In *atomic emission spectroscopy* the radiation source is the sample itself. The energy for

excitation of analyte atoms is supplied by a plasma, a flame, an oven, or an electric arc or

spark. The signal is the measured intensity of the source at the wavelength of interest. In

atomic absorption spectroscopy the radiation source is usually a line source such as a

hollow cathode lamp, and the signal is the absorbance. The latter is calculated from the

radiant power of the source and the resulting power after the radiation has passed through

the atomized sample. In *atomic fluorescence spectroscopy*, an external radiation source

is used, and the fluorescence emitted, usually at right angles to the source, is measured.

The signal is the intensity of the fluorescence emitted.

28-2.   **(a)** *Atomization* is a process in which a sample, often in solution, is volatilized and

decomposed to form an atomic vapor.

(c) *Doppler broadening* is an increase in the width of the atomic lines caused by the

Doppler effect in which atoms moving toward a detector absorb or emit wavelengths that

are slightly shorter than those absorbed or emitted b atoms moving at right angles to the

detector. The reverse effect is observed for atoms moving away from the detector.

(e) A *plasma* is a conducting gas that contains a large concentration of ions and/or

electrons.

(g) A *hollow cathode lamp* consists of a tungsten wire anode and a cylindrical cathode

sealed in a glass tube that contains argon at a pressure of 1 to 5 torr. The cathode is

constructed from or supports the element whose emission spectrum is desired.

(i) An *additive interference*, also called a blank interference, produces an effect that is independent of the analyte concentration. It could be eliminated with a perfect blank solution.

(k) A *chemical interference* in atomic spectroscopy is encountered when a species interacts with the analyte in such a way as to alter the spectral emission or absorption characteristics of the analyte.

(m) A *protective agent* prevents interference by forming a stable, but volatile, compound with the analyte. It protects the analyte from forming non-volatile, but less stable interfering compounds.

28-3. In atomic emission spectroscopy, the analytical signal is produced by the relatively small number of *excited* atoms or ions, whereas in atomic absorption the signal results from absorption by the much larger number of *unexcited* species. Any small change in flame conditions dramatically influences the number of *excited species,* whereas such changes have a much smaller effect on the number of *unexcited species*.

28-4. In atomic absorption spectroscopy the source radiation is modulated to create an ac signal at the detector. The detector is made to reject the dc signal from the flame and measure the modulated signal from the source. In this way, background emission from the flame and atomic emission from the analyte is discriminated against and prevented from causing an interference effect.

28-5. The temperature and pressure in a hollow cathode lamp are much less than those in an ordinary flame. As a result, Doppler and collisional broadening effects are much less, and narrower lines results.

28-6. The temperatures are high which favors the formation of atoms and ions. Sample

residence times are long so that desolvation and vaporization are essentially complete.

The atoms and ions are formed in a nearly chemically inert environment. The high and

relatively constant electron concentration leads to fewer ionization interferences.

28-7. The radial geometry provides better stability and precision while the axial geometry can

achieve lower detection limits. Many ICP emission systems allow both geometries.

28-8. By linear interpolation

$$0.400 + (0.502 - 0.396)\frac{(0.600 - 0.400)}{(0.599 - 0.396)} = 0.504 \text{ ppm Pb}$$

28-9.

(b) $A_s = \dfrac{\varepsilon b V_s c_s}{V_t} + \dfrac{\varepsilon b V_x c_x}{V_t} = kV_s c_s + kV_x c_x$

(c) For the plot of A_s versus V_s, $A_s = mV_s + b$ where $m = kc_s$

and $b = kV_x c_x$

(e) From the values in the spreadsheet: $m = 0.00881$ and $b = 0.2022$

(g) From the values in the spreadsheet: $c_{Cu} = 28.0 \ (\pm 0.2) \text{ ppm}$

Chapter 29

29-1.　**(a)** The *Dalton* is one unified atomic mass unit and equal to 1/12 the mass of a neutral

$^{12}_{6}C$ atom.

(c) The *mass number* is the atomic or molecular mass expressed without units.

(e) In a *time-of-flight* analyzer ions with nearly the same kinetic energy traverse a field-

free region. The time required for an ion to reach a detector at the end of the field-free

region is inversely proportional to the mass of the ion.

29-2.　The ICP torch serves both as an atomizer and ionizer.

29-3.　Interferences fall into two categories: spectroscopic interferences and matrix

interferences. In a spectroscopic interference, the interfering species has the same mass-

to-charge ratio as the analyte. Matrix effects occur at high concentrations where

interfering species can interact chemically or physically to change the analyte signal.

29-4.　The higher resolution of the double focusing spectrometer allows the ions of interest to be

better separated from background ions than with a relative low resolution quadrupole

spectrometer. The higher signal-to-background ratio of the double focusing instrument

leads to lower detection limits than with the quadrupole instrument.

29-5.　The high energy of the beam of electrons used in EI sources is enough to break some

chemical bonds and produce fragment ions. Such fragment ions can be useful in

qualitative identification of molecular species.

29-6.　The ion selected by the first analyzer is called the precursor ion. It then undergoes

thermal decomposition, reaction with a collision gas, or photodecomposition to form

product ions that are analyzed by a second mass analyzer.

Chapter 30

30-1. **(a)** The *order of a reaction* is the numerical sum of the exponents of the concentration

terms in the rate law for the reaction.

(c) *Enzymes* are high molecular mass organic molecules that catalyze reactions of

biological importance.

(e) The *Michaelis constant* K_m is an equilibrium-like constant for the dissociation of the

enzyme-substrate complex. It is defined by the equation $K_m = (k_{-1} + k_2)/k_1$, where k_1 and

k_{-1} are the rate constants for the forward and reverse reactions in the formation of the

enzyme-substrate complex. The term k_2 is the rate constant for the dissociation of the

complex to give products.

(g) *Integral methods* use integrated forms of the rate equations to calculate

concentrations from kinetic data.

30-3. Advantages would include; (1) measurements are made relatively early in the reaction

before side reactions can occur; (2) measurements do not depend upon the determination

of absolute concentration but rather depend upon differences in concentration; (3)

selectivity is often enhanced in reaction-rate methods, particularly in enzyme-based

methods. Limitations would include; (1) lower sensitivity, since reaction is not allowed

to proceed to equilibrium; (2) greater dependence on conditions such as temperature,

ionic strength, pH and concentration of reagents; (3) lower precision since the analytical

signal is lower.

30-5. $[A]_t = [A]_0 e^{-kt}$ $\ln \dfrac{[A]_t}{[A]_0} = -kt$

For $t = t_{1/2}$, $[A]_t = [A]_0/2$ $\ln \dfrac{[A]_0/2}{[A]_0} = \ln (1/2) = -kt_{1/2}$

$\ln 2 = kt_{1/2}$

$t_{1/2} = \ln 2/k = 0.693/k$

30-6. (a) $\tau = 1/k = 1/0.497 \text{ s}^{-1} = 2.01 \text{ s}$

(c) $\ln \dfrac{[A]_0}{[A]_t} = kt$ $\tau = 1/k = t/\ln([A]_0/[A]_t) = 3876 \text{ s}/\ln(3.16/0.496) = 2.093 \times 10^3 \text{ s}$

(e) $t_{1/2} = 26.5 \text{ yr} \times \dfrac{365 \text{ d}}{1 \text{ yr}} \times \dfrac{24 \text{ h}}{1 \text{ d}} \times \dfrac{60 \text{ min}}{1 \text{ h}} \times \dfrac{60 \text{ s}}{1 \text{ min}} = 8.36 \times 10^8 \text{ s}$

$\tau = 1/k = t_{1/2}/0.693 = 8.36 \times 10^8 \text{ s}/0.693 = 1.2 \times 10^9 \text{ s}$

30-7. (a) $\ln \dfrac{[A]_t}{[A]_0} = -kt$ $k = -\dfrac{1}{t} \ln \dfrac{[A]_t}{[A]_0}$

$k = -\dfrac{1}{0.0100} \ln(0.75) = 28.8 \text{ s}^{-1}$

(c) $k = 0.288 \text{ s}^{-1}$

(e) $k = 1.07 \times 10^4 \text{ s}^{-1}$

30-8. Let m = no. half-lives $= \dfrac{t}{t_{1/2}} = \dfrac{-\dfrac{1}{k} \ln \dfrac{[A]}{[A]_0}}{-\dfrac{1}{k} \ln \dfrac{[A]_0/2}{[A]_0}}$

$m = \dfrac{\ln[A]/[A]_0}{\ln 1/2} = -1.4427 \ln([A]/[A]_0)$

(a) $m = -1.4427 \ln 0.90 = 0.152$

(c) $m = -1.4427 \ln 0.10 = 3.3$

198

(e) $m = -1.4427 \ln 0.001 = 10$

30-10. (a) $[R]_0 = 5.00[A]_0$ where 5.00 is the ratio of the initial reagent concentration to the initial concentration of the analyte.

At 1% reaction, $[A] = 0.99[A]_0$

$[R]_{1\%} = [R] - 0.01[A]_0 = 5.00[A]_0 - 0.01[A]_0 = 4.99[A]_0$

$Rate_{assumed} = k[R][A] = k(5.00[A]_0 \times 0.99[A]_0)$

$Rate_{true} = k(4.99[A]_0 \times 0.99[A]_0)$

relative error $= \dfrac{k(5.00[A]_0 \times 0.99[A]_0) - k(4.99[A]_0 \times 0.99[A]_0)}{k(4.99[A]_0 \times 0.99[A]_0)}$

$= \dfrac{(5.00 \times 0.99) - (4.99 \times 0.99)}{(4.99 \times 0.99)} = 0.00200$

relative error $\times 100\% = 0.2\%$

(c) $(50.00 - 49.99)/49.99 = 0.000200$ or 0.02%

(e) $(5.00 - 4.95)/4.95 = 0.0101$ or 1.0%

(g) $(100.00 - 99.95)/99.95 = 0.0005002$ or 0.05%

(i) $(10.000 - 9.368)/9.368 = 0.06746$ or 6.7%

(k) $(100.00 - 99.368)/99.368 = 0.00636$ or 0.64%

30-12. (a) Plot 1/Rate versus 1/[S] for known [S] to give a linear calibration curve. Measure rate for unknown [S], calculate 1/Rate and 1/[S]$_{unknown}$ from the working curve and find [S]$_{unknown}$.

(b) The intercept of the calibration curve is $1/v_{max}$ and the slope is K_m/v_{max}. Use the intercept to calculate $K_m = $ slope/intercept, and $v_{max} = 1/$intercept.

30-13.

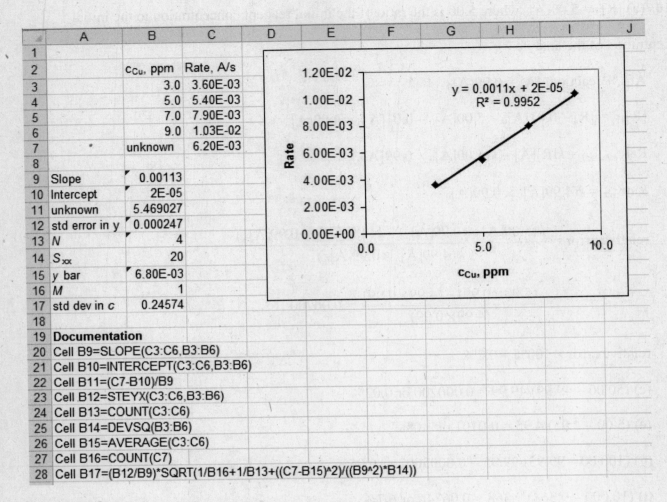

	A	B	C	D	E	F	G	H	I	J
1										
2		c_{Cu}, ppm	Rate, A/s							
3		3.0	3.60E-03							
4		5.0	5.40E-03							
5		7.0	7.90E-03							
6		9.0	1.03E-02							
7		unknown	6.20E-03							
8										
9	Slope	0.00113								
10	Intercept	2E-05								
11	unknown	5.469027								
12	std error in y	0.000247								
13	N	4								
14	S_{xx}	20								
15	y bar	6.80E-03								
16	M	1								
17	std dev in c	0.24574								
18										
19	**Documentation**									
20	Cell B9=SLOPE(C3:C6,B3:B6)									
21	Cell B10=INTERCEPT(C3:C6,B3:B6)									
22	Cell B11=(C7-B10)/B9									
23	Cell B12=STEYX(C3:C6,B3:B6)									
24	Cell B13=COUNT(C3:C6)									
25	Cell B14=DEVSQ(B3:B6)									
26	Cell B15=AVERAGE(C3:C6)									
27	Cell B16=COUNT(C7)									
28	Cell B17=(B12/B9)*SQRT(1/B16+1/B13+((C7-B15)^2)/((B9^2)*B14))									

We report the concentration of the unknown as 5.5 ± 0.2 ppm

30-15. $$\text{Rate} = R = \frac{k_2[E]_0[tryp]_t}{[tryp]_t + K_m}$$

Assume $K_m \gg [tryp]_t$

$$R = \frac{v_{max}[tryp]_t}{K_m} \qquad \text{and} \qquad [tryp]_t = K_m / v_{max}$$

$[tryp]_t = (0.18 \ \mu M/min)(4.0 \times 10^{-4} \ M)/(1.6 \times 10^{-3} \ \mu M/min) = 0.045 \ M$

200

30-17.

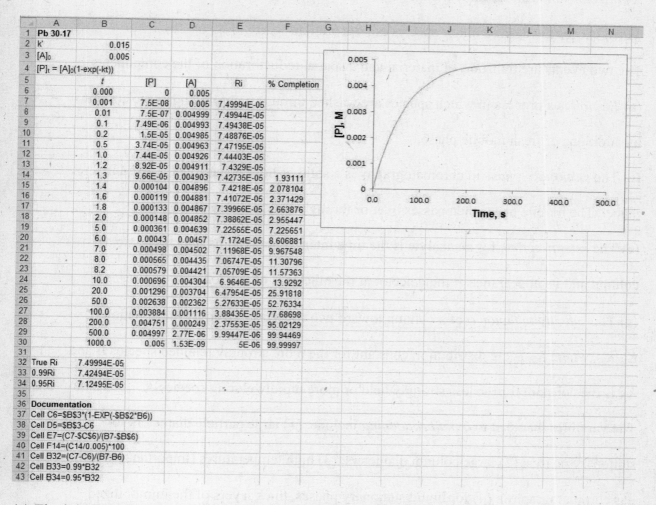

	A	B	C	D	E	F
1	Pb 30-17					
2	k'	0.015				
3	[A]₀	0.005				
4	[P]ₜ = [A]₀(1-exp(-kt))					
5		t	[P]	[A]	Ri	% Completion
6		0.000	0	0.005		
7		0.001	7.5E-08	0.005	7.49994E-05	
8		0.01	7.5E-07	0.004999	7.49944E-05	
9		0.1	7.49E-06	0.004993	7.49438E-05	
10		0.2	1.5E-05	0.004985	7.48876E-05	
11		0.5	3.74E-05	0.004963	7.47195E-05	
12		1.0	7.44E-05	0.004926	7.44403E-05	
13		1.2	8.92E-05	0.004911	7.4329E-05	
14		1.3	9.66E-05	0.004903	7.42735E-05	1.93111
15		1.4	0.000104	0.004896	7.4218E-05	2.078104
16		1.6	0.000119	0.004881	7.41072E-05	2.371429
17		1.8	0.000133	0.004867	7.39966E-05	2.663876
18		2.0	0.000148	0.004852	7.38862E-05	2.955447
19		5.0	0.000361	0.004639	7.22565E-05	7.225651
20		6.0	0.00043	0.00457	7.1724E-05	8.606881
21		7.0	0.000498	0.004502	7.11968E-05	9.967548
22		8.0	0.000565	0.004435	7.06747E-05	11.30796
23		8.2	0.000579	0.004421	7.05709E-05	11.57363
24		10.0	0.000696	0.004304	6.9646E-05	13.9292
25		20.0	0.001296	0.003704	6.47954E-05	25.91818
26		50.0	0.002638	0.002362	5.27633E-05	52.76334
27		100.0	0.003884	0.001116	3.88435E-05	77.68698
28		200.0	0.004751	0.000249	2.37553E-05	95.02129
29		500.0	0.004997	2.77E-06	9.99447E-06	99.94469
30		1000.0	0.005	1.53E-09	5E-06	99.99997
31						
32	True Ri	7.49994E-05				
33	0.99Ri	7.42494E-05				
34	0.95Ri	7.12495E-05				
35						
36	**Documentation**					
37	Cell C6=$B$3*(1-EXP(-$B$2*B6))					
38	Cell D5=$B$3-C6					
39	Cell E7=(C7-$C$6)/(B7-$B$6)					
40	Cell F14=(C14/0.005)*100					
41	Cell B32=(C7-C6)/(B7-B6)					
42	Cell B33=0.99*B32					
43	Cell B34=0.95*B32					

(a) The initial rate drops to $0.99R_i$ between times 1.3 and 1.4 s, which is $\approx$ 2% of the reaction.

(b) Between 6.0 and 7.0 s so a little over 9% of the reaction is completed.

Chapter 31

31-1. A *collector ion* is an ion added to a solution that forms a precipitate with the reagent which carries the desired minor species out of solution.

31-3. The two events are transport of material and a spatial redistribution of the components.

31-5. **(a)** *Elution* is a process in which species are washed through a chromatographic column by additions of fresh mobile phase.

(c) The *stationary phase* in chromatography is a solid or liquid phase that is fixed in place. The mobile phase then passes over or through the stationary phase.

(e) The *retention time* for an analyte is the time interval between its injection onto a column and its appearance at the detector at the other end of the column.

(g) The *selectivity factor* α of a column toward two species is given by the equation $\alpha = K_B/K_A$, where K_B is the distribution constant for the more strongly retained species B and K_A is the constant for the less strongly held or more rapidly eluting species A.

31-7. The variables that lead to *band broadening* include: (1) large particle diameters for stationary phases; (2) large column diameters; (3) high temperatures (important only in gas chromatography); (4) for liquid stationary phases, thick layers of the immobilized liquid; and (5) very rapid or very slow flow rates.

31-9. Determine the retention time t_R for a solute and the width of the solute peak at its base, W. The number of plates N is then $N = 16(t_R/W)^2$.

31-11. $[X]_i = \left(\dfrac{V_{aq}}{V_{org}K + V_{aq}} \right)^i [X]_0$

(a) $[X]_1 = \left(\dfrac{50.0}{40.0 \times 8.9 + 50.0} \right)(0.200) = 0.0246 \text{ M}$

(b) $[X]_2 = \left(\dfrac{50.0}{20.0 \times 8.9 + 50.0}\right)^2 (0.200) = 9.62 \times 10^{-3}$ M

(c) $[X]_4 = \left(\dfrac{50.0}{10.0 \times 8.9 + 50.0}\right)^4 (0.200) = 3.35 \times 10^{-3}$ M

(d) $[X]_8 = \left(\dfrac{50.0}{5.0 \times 8.9 + 50.0}\right)^8 (0.200) = 1.23 \times 10^{-3}$ M

31-13. $[A]_i = \left(\dfrac{V_{aq}}{V_{org}K + V_{aq}}\right)^i [A]_0$ $i = \dfrac{\log\left([A]_i / [A]_0\right)}{\log\left(\dfrac{V_{aq}}{V_{org}K + V_{aq}}\right)}$

(a) $i = \dfrac{\log\left(\dfrac{1.00 \times 10^{-4}}{0.0500}\right)}{\log\left(\dfrac{25.0}{25.0 \times 8.9 + 25.0}\right)} = 2.7$ extractions. So 3 extractions are needed.

The total volume would be 75 mL with 3 extractions.

(b) As in part (a), $i = 4.09$ extractions, so 5 extractions are needed.

The total volume would be 5×10 mL $= 50$ mL

(c) $i = 11.6$ so 12 extractions are needed

The total volume would be 12×2 mL $= 24$ mL

31.15. If 99% of the solute is removed then 1% of solute remains and $[A]_i / [A]_0 = 0.01$.

(a) $\dfrac{[A]_i}{[A]_0} = \left(\dfrac{50.0}{25.0K + 50.0}\right)^2 = 0.01$

$\quad (0.01)^{1/2}(25.0K + 50.0) = 50.0$

$\quad 2.5K + 5.0 = 50.0$

$\quad K = (50.0 - 5.0)/2.5 = 18.0$

(b) $\dfrac{[A]_i}{[A]_0} = \left(\dfrac{50.0}{10.0K + 50.0}\right)^5 = 0.01$

$(0.01)^{1/5}(10.0K + 50.0) = 50.0$

$3.98K + 19.9 = 50.0$

$K = (50.0 - 19.9)/3.98 = 7.56$

31-16. (a) If 1.00×10^{-4} % of the solute remains, $[A]_i / [A]_0 = 1.00 \times 10^{-6}$.

$\dfrac{[A]_i}{[A]_0} = \left(\dfrac{30.0}{10.0K + 30.0}\right)^4 = 1.00 \times 10^{-6}$

$(1 \times 10^{-6})^{1/4}(10.0K + 30.0) = 30.0$

$0.316K + 0.949 = 30.0$

$K = (30.0 - 0.949)/0.31 = 91.9$

$K = (30.0 - 3.00)/1.00 = 27.0$

31-17. (a) Recognizing that in each of the solutions [HA] = 0.0750 due to dilution, from the data

for solution 1,

$[HA]_{org} = 0.0454$ M

$[HA]_{aq} = \dfrac{25.0(0.0750) - 25.0(0.0454)}{25.0} = 0.0296$ M

$K = [HA]_{org}/[HA]_{aq} = 0.0454/0.0296 = 1.53$

(b) For solution 3, after extraction

$[HA]_{aq} = [HA]_{org} / K = 0.0225 / 1.53 = 0.0147$ M

$[A^-] = (\text{mols } HA_{tot} - \text{mols } HA_{aq} - \text{mols } HA_{org})/(25.0 \text{ mL})$

$[A^-] = \dfrac{(25.0)(0.0750) - (25.0)(0.0147) - (25.0)(0.0225)}{25.0} = 0.0378$ M

(c) Since $[H^+] = [A^-]$, $K_a = (0.0378)^2/(0.0147) = 0.0972$

31-19. (a) amount H$^+$ resulting from exchange = 15.3 mL × 0.0202 mmol/mL = 0.3091 mmol

mmols H$^+$ = mol cation = 0.3091 in 0.0250 L sample

0.3091 mmol cation/0.0250 L = 12.36 mmol cation/L

(b) $\dfrac{12.36 \text{ mmol cation}}{\text{L}} \times \dfrac{1 \text{ mmol CaCO}_3}{2 \text{ mmol cation}} \times \dfrac{100.087 \text{ mg CaCO}_3}{\text{mmol CaCO}_3} = 619 \text{ mg CaCO}_3/\text{L}$

31-21. [HCl] = $17.53 \text{ mL} \times \dfrac{0.02932 \text{ mmol NaOH}}{\text{mL}} \times \dfrac{1 \text{ mmol HCl}}{1 \text{ mmol NaOH}} \times \dfrac{1}{25.00 \text{ mL}}$

= 0.02056 mmol/mL

amount H$_3$O$^+$/mL from exchange = 35.94 mL×0.02932 mmol/mL/10.00 mL = 0.10538

= (no. mmol HCl + 2 × no. mmol MgCl$_2$)/mL

mmol MgCl$_2$/mL = (0.10536 − 0.02056)/2 = 0.0424

The solution is thus 0.02056 M in HCl and 0.0424 M in MgCl$_2$.

31-23. From equation 31-13,

$$u_0 = F/\varepsilon\pi r^2 = F/\varepsilon\pi(d/2)^2 = \dfrac{48 \text{ cm}^3/\text{min}}{0.43 \times 3.1415 \times \left(\dfrac{0.50 \text{ cm}}{2}\right)^2}\left(\dfrac{1 \text{ min}}{60 \text{ s}}\right) = 9.5 \text{ cm/s}$$

31-25. (a) $k = (t_R - t_M)/t_M$

For A, $k_A = (5.4 - 3.1)/3.1 = 0.742 = 0.74$

For B, $k_B = (13.3 - 3.1)/3.1 = 3.29 = 3.3$

For C, $k_C = (14.1 - 3.1)/3.1 = 3.55 = 3.5$

For D, $k_D = (21.6 - 3.1)/3.1 = 5.97 = 6.0$

(b) $K = k\,V_M/V_S$

For A, $K_A = 0.742 \times 1.37 \,/\, 0.164 = 6.2$

For compound B, $K_B = 3.29 \times 1.37\ 0.164 = 27$

For compound C, $K_C = 3.55 \times 1.37/0.164 = 30$

For compound D, $K_D = 5.97 \times 1.37/0.164 = 50$

Problems 31-28 through 31-31: See next two spreadsheets

	A	B	C	D	E
1	**Problem 31-28**				
2	Compound	t_R, min	W	N	
3	Air	1.9			
4	Methylcyclohexane	10	0.76	2770.083	
5	Methylcyclohexene	10.9	0.82	2827.127	
6	Toluene	13.4	1.06	2556.924	
7					
8	Average N			2718.045	
9	Std. Dev.			142.4196	
10	Column Length, L			40	
11	Plate Height, H			0.014716	
12	**Spreadsheet Documentaion**				
13	Cell D4=16*(B4/C4)^2				
14	Cell D8=AVERAGE(D4:D6)				
15	Cell D9=STDEV.S(D4:D6)				
16	Cell D11=D10/D8				
17					
18	**Problem 31-29**				
19	R_s (methylcylohexene - methyl cyclohexane)				1.14
20	R_s (methylcyclohexene - toluene)				2.66
21	R_s (toluene - methylcylohexane)				3.74
22	**Spreadsheet Documentation**				
23	Cell E19=2*(B5-B4)/(C5+C4)				
24	Cell E21=2*(B6-B4)/(C4+C6)				
25					
26	**Problem 31-30**				
27	To obtain R_s = 1.75	N_2	6413.6		
28	Column Length, L		94.38549		
29	Retention time t_R		25.72005		
30	**Spreadsheet Documentation**				
31	Cell C27=D8*1.75^2/E19^2				
32	Cell C28=C27*D11				
33	Cell C29=B5*1.75^2/E19^2				

The following spreadsheet is a continuation of the previous spreadsheet.

	A	B	C	D	E
35	**Problem 31-31**				
36	k (methylcyclohexane)				4.263158
37	k (methylcyclohexene)				4.736842
38	k (toluene)				6.052632
39	V_M				62.6
40	V_s				19.6
41	K (methylcyclohexane)				13.62
42	K (methylcyclohexene)				15.13
43	K (toluene)				19.33
44	α (methylcyclohexane-methylcyclohexene)				1.11
45	**Spreadsheet Documentation**				
46	Cell E36=(B4-$B$3)/$B$3				
47	Cell E41=E36*$E$39/$E$40				
48	Cell E44=(B5-B3)/(B4-B3)				

Problems 31-32 and 31-33

	A	B	C	D	E	F	G
1	**Problem 31-32**			**Problem 31-33**			
2							
3	K (M)	5.99		5.81			
4	K (N)	6.16		6.20			
5	R	1.5					
6	V_s/V_M	0.425					
7	H	1.50E-03					
8	F	6.75					
9							
10	k (M)	2.54575		2.54575			
11	k (N)	2.618		2.618			
12	α	1.028381		1.067126			
13	N	90274.26		17376.19			
14	L	135.4114		26.06428			
15	$(t_R)_N$	72.5805		13.97046			
16	**Spreadsheet Documentation**						
17	Cell B10=$B$3*$B$6						
18	Cell B11=$B$4*$B$6						
19	Cell B12=B4/B3						
20	Cell B13=16*$B$5^2*(B12/(B12-1))^2*((1+B11)/B11)^2						
21	Cell B14=B13*$B$7						
22	Cell B15=(16*$B$5^2*$B$7/$B$8)*(B12/(B12-1))^2*(1+B11)^3/B11^2						

Chapter 32

32-1. In *gas-liquid chromatography*, the stationary phase is a liquid that is immobilized on a

solid. Retention of sample constituents involves equilibria between a gaseous and a

liquid phase. In *gas-solid chromatography*, the stationary phase is a solid surface that

retains analytes by physical adsorption. Here separation involves adsorption equilibria.

32-3. Gas-solid chromatography is used primarily for separating low molecular mass gaseous

species, such as carbon dioxide, carbon monoxide and oxides of nitrogen.

32-5. A chromatogram is a plot of detector response versus time. The peak position, retention

time, can reveal the identity of the compound eluting. The peak area is related to the

concentration of the compound.

32-7. In *open tubular or capillary columns,* the stationary phase is held on the inner surface of

a capillary, whereas in *packed columns,* the stationary phase is supported on particles that

are contained in a glass or metal tube. Open tubular columns contain an enormous

number of plates that permit rapid separations of closely related species. They suffer

from small sample capacities.

32-9. Sample injection volume, carrier gas flow rate and column condition are among the

parameters which must be controlled for highest precision quantitative GC. The use of

an internal standard can minimize the impact of variations in these parameters.

32-11. (a) Advantages of thermal conductivity: general applicability, large linear range,

simplicity, nondestructive.

Disadvantage: low sensitivity.

(b) Advantages of flame ionization: high sensitivity, large linear range, low noise,

ruggedness, ease of use, and response that is largely independent of flow rate.

Disadvantage: destructive.

(c) Advantages of electron capture: high sensitivity selectivity towards halogen-

containing compounds and several others, nondestructive.

Disadvantage: small linear range.

(d) Advantages of thermionic detector: high sensitivity for compounds containing

nitrogen and phosphorus, good linear range.

Disadvantages: destructive, not applicable for many analytes.

(e) Advantages of photoionization: versatility, nondestructive, large linear range.

Disadvantages: not widely available, expensive.

32-13. Megabore columns are open tubular columns that have a greater inside diameter (530

μm) than typical open tubular columns (150 to 320 μm). Megabore columns can tolerate

sample sizes similar to those for packed columns, but with significantly improved

performance characteristics. Thus, megabore columns can be used for preparative scale

GC purification of mixtures where the compound of interest is to be collected and further

analyzed using other analytical techniques.

32-15. Currently, liquid stationary phases are generally bonded and/or cross-linked in order to

provide thermal stability and a more permanent stationary phase that will not leach off

the column. Bonding involves attaching a monomolecular layer of the stationary phase to

the packing surface by means of chemical bonds. Cross linking involves treating the

stationary phase while it is in the column with a chemical reagent that creates cross links

between the molecules making up the stationary phase.

32-17. Fused silica columns have greater physical strength and flexibility than glass open tubular

columns and are less reactive toward analytes than either glass or metal columns.

32-19. (a) Band broadening arises from very high or very low flow rates, large particles making

up packing, thick layers of stationary phase, low temperature, and slow injection rates.

(b) Band separation is enhanced by maintaining conditions so that k lies in the range of 1

to 10, using small particles for packing, limiting the amount of stationary phase so that

particle coatings are thin, and injecting the sample rapidly.

32-21.

	A	B	C	D	E
1	Problem 32-21				
2	Compound	Relative area	Correction factor	Corrected area	Percentage
3	A	32.5	0.70	46.428571	21.09
4	B	20.7	0.72	28.750000	13.06
5	C	60.1	0.75	80.133333	36.40
6	D	30.2	0.73	41.369863	18.79
7	E	18.3	0.78	23.461538	10.66
8					
9			Total area	220.143306	
10					
11	Spreadsheet Documentation				
12	Cell D3=B3/C3				
13	Cell D9=SUM(D3:D7)				
14	Cell E3=D3/$D$9*100				

Chapter 33

33-1. **(a)** Substances that are somewhat volatile and are thermally stable.

(c) Substances that are ionic.

(e) High molecular mass compounds that are soluble in nonpolar solvents.

(g) Chiral compounds (enantiomers).

33-2. **(a)** In an *isocratic elution,* the solvent composition is held constant throughout the

elution.

(c) In a *normal-phase packing,* the stationary phase is quite polar and the mobile phase is

relatively nonpolar.

(e) In a *bonded-phase packing*, the stationary phase liquid is held in place by chemically

bonding it to the solid support.

(g) In *ion-pair chromatography* a large organic counter-ion is added to the mobile phase

as an ion-pairing reagent. Separation is achieved either through partitioning of the

neutral ion-pair or as a result of electrostatic interactions between the ions in solution and

charges on the stationary phase resulting from adsorption of the organic counter-ion.

(i) *Gel filtration* is a type of size-exclusion chromatography in which the packings are

hydrophilic, and eluents are aqueous. It is used for separating high molecular mass polar

compounds.

33-3. **(a)** diethyl ether, benzene, *n*-hexane.

33-4. **(a)** ethyl acetate, dimethylamine, acetic acid.

33-5. In *adsorption chromatography,* separations are based on adsorption equilibria between

the components of the sample and a solid surface. In *partition chromatography,*

separations are based on distribution equilibria between two immiscible liquids.

33-7. *Gel filtration* is a type of size-exclusion chromatography in which the packings are hydrophilic and eluents are aqueous. It is used for separating high molecular mass polar compounds. *Gel permeation chromatography* is a type of size-exclusion chromatography in which the packings are hydrophobic and the eluents are nonaqueous. It is used for separating high molecular mass nonpolar species.

33-9. In an *isocratic elution,* the solvent composition is held constant throughout the elution. Isocratic elution works well for many types of samples and is simplest to implement. In a *gradient elution,* two or more solvents are employed and the composition of the eluent is changed continuously or in steps as the separation proceeds. Gradient elution is best used for samples in which there are some compounds separated well and others with inordinately long retention times.

33-11. In *suppressor-column ion chromatography* the chromatographic column is followed by a column whose purpose is to convert the ions used for elution to molecular species that are largely nonionic and thus do not interfere with conductometric detection of the analyte species. In *single-column ion chromatography*, low capacity ion exchangers are used so that the concentrations of ions in the eluting solution can be kept low. Detection then is based on the small differences in conductivity caused by the presence of eluted sample components.

33-13. Comparison of Table 33-1 with Table 32-1 suggests that the GC detectors that are suitable for HPLC are the mass spectrometer, FTIR and possible photoionization. Many of the GC detectors are unsuitable for HPLC because they require the eluting analyte components to be in the gas-phase.

33-15. A number of factors that influence separation are clearly temperature dependent

including distribution constants and diffusion rates. In addition, temperature changes can

influence selectivity if components A and B are influenced differently by changes in

temperature. Because resolution depends on all these factors, resolution will also be

temperature dependent.

(a) For a reversed phase chromatographic separation of a steroid mixture, selectivity and,

as a consequence, separation could be influenced by temperature dependent changes in

distribution coefficients.

(b) For an adsorption chromatographic separation of a mixture of isomers, selectivity

and, as a consequence, separation could be influenced by temperature dependent changes

in distribution coefficients.

Chapter 34

34-1. **(a)** Nonvolatile or thermally unstable species that contain no chromophoric groups.

(c) Inorganic anions and cations, amino acids, catecholamines, drugs, vitamins, carbohydrates, peptides, proteins, nucleic acids, nucleotides, and polynucleotides.

(e) Proteins, synthetic polymers, and colloidal particles.

34-2. **(a)** A *supercritical fluid* is a substance that is maintained above its critical temperature so that it cannot be condensed into a liquid no matter how great the pressure.

(c) In *two-dimensional thin layer chromatography,* development is carried out with two solvents that are applied successively at right angles to one another.

(e) The *critical micelle concentration* is the level above which surfactant molecules begin to form spherical aggregates made up to 40 to 100 ions with their hydrocarbon tails in the interior of the aggregate and their charged ends exposed to water on the outside.

34-3. The properties of a supercritical fluid that are important in chromatography include its density, its viscosity, and the rates at which solutes diffuse in it. The magnitude of each of these lies intermediate between a typical gas and a typical liquid.

34-5. (a) Instruments for supercritical-fluid chromatography are very similar to those for HPLC except that in SFC there are provisions for controlling and measuring the column pressure. (b) SFC instruments differ substantially from those used for GC in that SFC instruments must be capable of operating at much higher mobile phase pressures than are typically encountered in GC'

34-7. Their ability to dissolve large nonvolatile molecules, such as large *n*-alkanes and polycyclic aromatic hydrocarbons.

34-9. **(a)** An increase in flow rate results in a decrease in retention time.

(b) An increase in pressure results in a decrease in retention time.

(c) An increase in temperature results in a decrease in density of supercritical fluids and

thus an increase in retention time.

34-11. *Electroosmotic flow* is the migration of the solvent towards the cathode in an

electrophoretic separation. This flow is due to the electrical double layer that develops at

the silica/solution interface. At pH values higher than 3 the inside wall of the silica

capillary becomes negatively charged leading to a build-up of buffer cations in the

electrical double layer adjacent to the wall. The cations in this double layer are attracted

to the cathode and, since they are solvated they drag the bulk solvent along with them.

34-13. Under the influence of an electric field, mobile ions in solution are attracted or repelled

by the negative potential of one of the electrodes. The rate of movement toward or away

from a negative electrode is dependent on the net charge on the analyte and the size and

shape of analyte molecules. These properties vary from species to species. Hence, the

rate at which molecules migrate under the influence of the electric field vary, and the

time it takes them to traverse the capillary varies, making separations possible.

34-15. The electrophoretic mobility is given by

$$v = \frac{\mu_e V}{L} = \frac{5.13 \times 10^{-4} \text{ cm}^2 \text{ s}^{-1} \text{ V}^{-1} \times 20000 \text{ V}}{50} = 0.2052 \text{ cm s}^{-1}$$

The electroosmotic flow rate is given as 0.65 mm s^{-1} = 0.065 cm s^{-1}

Thus, the total flow rate = 0.2052 + 0.065 = 0.2702 cm s^{-1}, and

$t = [(40.0 \text{ cm})/0.2702 \text{ cm s}^{-1})] \times (1 \text{ min}/60 \text{ s}) = 2.5 \text{ min}$

34-17. Higher column efficiencies and the ease with which pseudostationary phase can be

altered.

34-19. B^+ followed by A^{2+} followed by C^{3+}.